M381 Number Theory and
Mathematical Logic

GW00598012

Mathematical Logic **Unit 1**

Computability

Prepared by the Course Team

The M381 Mathematical Logic Course Team

The Mathematical Logic half of the course was produced by the following team:

Roberta Cheriyan	*Course Manager*
Derek Goldrei	*Course Team Chair* and *Academic Editor*
Jeremy Gray	*History Consultant*
Mary Jones	*Critical Reader*
Roger Lowry	*Publishing Editor*
Alan Pears	*Author*
Alan Slomson	*Author*
Frances Williams	*Critical Reader*

with valuable assistance from:

The Maths Production Unit, Learning & Teaching Solutions:
Becky Browne, Jim Campbell, Nicky Kempton, Bill Norman, Sharon Powell, Katie Sayce, Penny Tee

Alison Cadle	*TEX Consultant*
Michael Goldrei	*Cover Design Consultant*
Vicki McCulloch	*Cover Designer*

The external assessor was:

Jeff Paris	*Professor of Pure Mathematics, University of Manchester*

The Course Team would like to acknowledge their reliance on the previous work of Alan Slomson and of Alex Wilkie, Professor of Mathematical Logic, University of Oxford.

This publication forms part of an Open University course. Details of this and other Open University courses can be obtained from the Student Registration and Enquiry Service, The Open University, PO Box 197, Milton Keynes, MK7 6BJ, United Kingdom: tel. +44 (0)870 300 6090, e-mail general-enquiries@open.ac.uk

Alternatively, you may visit the Open University website at http://www.open.ac.uk where you can learn more about the wide range of courses and packs offered at all levels by The Open University.

To purchase a selection of Open University course materials, visit http://www.ouw.co.uk, or contact Open University Worldwide, Michael Young Building, Walton Hall, Milton Keynes, MK7 6AA, United Kingdom, for a brochure: tel. +44 (0)1908 858793, fax +44 (0)1908 858787, e-mail ouw-customer-services@open.ac.uk

The Open University, Walton Hall, Milton Keynes, MK7 6AA.

First published 2003. Reprinted as new edition 2007, with corrections.

Edited, designed and typeset by The Open University, using the Open University TEX System.

Printed in the United Kingdom by The Charlesworth Group, Wakefield.

ISBN 978 0 7492 2280 2

3.1

CONTENTS

INTRODUCTION

In the Mathematical Logic half of the course we shall be studying two main topics: *mathematical logic* and *computability*.

Logic concerns itself with sound methods of reasoning. It can be made mathematical in two different ways: by restricting its content to methods of reasoning used in mathematics and by using mathematical methods to study logic. In *mathematical logic* these come together. We shall concern ourselves with reasoning used in mathematics and we shall study this in a mathematical way.

Computability deals with fundamental theoretical questions about algorithms and computers.

At first sight there is no direct connection between computability and mathematical logic but we shall see that they are intimately related, both through the historical origins of these subjects and through the techniques which they use.

In studying these topics we shall meet ideas that have a wide range of applicability — in computer science, philosophy and linguistics as well as in mathematics itself. However, our study of them is directed towards one main goal: Gödel's celebrated Incompleteness Theorems.

Kurt Gödel (1906–1978) proved the first of these theorems in 1931.

Historical beginning

Logic has ancient roots. The systematic study of logical reasoning goes back at least as far as Aristotle, who died some 2300 years ago. In the space at our disposal in this Introduction, we can give only a brief, highly selective and simplified account of the development of the logical and other ideas which led to the mathematics discussed in the Mathematical Logic part of this course.

The mechanization of knowledge

We attribute to Leibniz the idea of making knowledge mechanical. He had the idea that a scientifically designed language would help people to think clearly and would make reasoning easy by providing a mechanical method for drawing conclusions. He thought that when such a language had been perfected, people desiring to settle a controversy on any subject would need only to pick up their pens and say 'let us calculate'. He was probably influenced by Descartes's invention of analytic geometry, which converts hard geometrical problems into more routine algebraic calculations, and by the great success of the differential and integral calculus, which provides many algorithmic methods for solving problems about tangents and areas (and in the development of which Leibniz played an important role).

Although Leibniz sketched out his ideas more than once, there was little further development until the work of Boole (George Boole, 1815–1864) who in 1847 published *The Mathematical Analysis of Logic, being an essay towards a calculus of deductive reasoning*. Boole's main idea was that it is possible to have an algebra of entities that are not in any sense numbers, and his book shows how to develop an algebra of propositions, which, in a slightly different form, we now call *Boolean algebra*. The subsequent work of Frege (Gottlob Frege, 1848–1925), Russell (Bertrand Russell, 1872–1970), Whitehead (Alfred North Whitehead, 1861–1947) and others produced a system of logic adequate for handling most of mathematics. In this sense they achieved Leibniz's dream of a universal language, at least as far as mathematics is concerned.

Gottfried Wilhelm Leibniz, 1646–1716, was the most distinguished mathematician and philosopher of his generation, co-inventor of the calculus independently of Isaac Newton. In *Unit 8* we shall provide more biographical details of Leibniz and some of the other figures mentioned here. (Picture © George Bernard/Science Photo Library)

Although this work resulted in a universal language for most of mathematics, it left open the question as to whether, using this language, all mathematical problems could be solved by mechanical manipulation, thus fully realizing Leibniz's idea. One of the main purposes of this course is to answer this question. In fact we answer the question in relation to just part of mathematics, that dealing with the theory of numbers, but the answer to this restricted question throws light on the whole of Leibniz's project. We thus call the question we address *Leibniz's Question* and we formulate it as follows.

Leibniz's Question

Is there an algorithm for deciding which statements of number theory are true?

To answer this question we shall need to study both the notion of an *algorithm*, which we do in *Units 1* to *3*, and a *formal language* for number theory, which we discuss in *Units 4* to *7*. We shall give an answer to this question in *Unit 8*.

The consistency of mathematics

Frege's logical calculus arose from his interest in philosophical problems about the nature of mathematics. He came to the view that mathematics, or at least the part of it dealing with numbers and their arithmetic, was nothing other than logic. He meant by this that all the basic concepts could be defined in terms of logical notions and their theory could be developed using only logical principles. To justify this claim he needed to be explicit not only about his starting hypotheses but also about the principles used to deduce conclusions from them. His formal logical language was introduced, as he explained, 'to provide us with the most reliable test of the validity of a chain of inferences and to point out every presupposition that tries to sneak in unnoticed' (*Begriffsschrift*, 1879). *Units 4* to *6* of this course describe a logical calculus based on Frege's ideas.

At about the same time Cantor (Georg Cantor, 1845–1918) was developing his theory of mathematical infinity arising from his work on the convergence of Fourier series. The long-standing tradition in mathematics, deriving from Aristotle, was that we can consider important objects like the natural numbers only as *potentially infinite*, meaning that however (finitely) many numbers one might list, there is always an extra one which can be added to the list. But mathematicians were wary of treating such a list, which nowadays we are happy to call the set of natural numbers, as a single object, described as being *actually infinite*. Cantor, however, developed a fruitful theory of actually infinite sets. Some of his ideas are explained later in this course, in *Unit 2*.

Russell was also interested in the philosophy of mathematics and he arrived, independently, at a view similar to that of Frege. He began studying Frege's work in 1902, and subsequently with Whitehead he published *Principia Mathematica* in which it is shown in detail how a great deal of mathematics can be derived within a logical calculus.

In studying Cantor's work on infinite sets Russell discovered a seeming contradiction, known now as *Russell's Paradox*, and he noted that this contradiction could be derived in Frege's logical system. This was a defining moment in set theory and mathematical logic. There were varied reactions to Russell's Paradox. For Frege and Russell it was a great blow to their concept of mathematics as logic. Frege eventually abandoned this view and began to try to base mathematics on geometrical ideas. Russell put forward a number of different ways of avoiding his paradox but, however ingenious his ideas were, the effect was considerably to reduce the plausibility of the theory that mathematics could be based on logic and nothing more.

Cantor's work had already met with some suspicion because it was contrary to the Aristotelian tradition. The discovery of Russell's Paradox only reinforced this reaction among some mathematicians. Cantor, who had noticed some similar problems with his theory before Russell's Paradox was published, was not greatly concerned about the paradox and it has been argued subsequently that this is because the paradox applies only to the Frege–Russell concept of a set and not to Cantor's.

The mathematician Hilbert was in the middle of this debate. He was interested in the foundations of mathematics and was a great admirer of Cantor's work. He noted also that, independently of Cantor's work, the actually infinite was involved in the work of Weierstrass and others in establishing mathematical analysis (the theory that underlies the differential and integral calculus) on a firm foundation, since real numbers are constructed as actually infinite objects (for example, as the limits of infinite sequences of rational numbers or as infinite decimal expansions). Hilbert's idea was to show that these uses of infinite sets are free from contradiction, that is *consistent*, in the following way. Frege, Russell and Whitehead had shown that mathematical analysis could be derived within a logical calculus. This calculus involved formulas expressing mathematical propositions together with rules for deriving theorems from an initial set of axioms. Although these formulas are interpreted as being about infinite objects, viewed syntactically they are nothing other than finite strings of symbols. For example, the formula

$$\forall x, y \in \mathbb{R} \; (x + y) = (y + x)$$

expresses the fact that addition of real numbers is commutative. Thus it is interpreted as a statement about infinitely many different real numbers. But viewed syntactically it is nothing more than a string of 17 symbols.

The rules of the logical calculus, as we shall see in *Units 5* and *6*, enable us to manipulate these formulas in a mechanical way without any regard to their meaning. This gave Hilbert hope that it would be possible to prove, using arguments about the formulas considered just as strings of symbols, that the manipulations allowed by the system did not lead to a contradiction. This would be reasoning about finite objects. *Finitary reasoning* of this kind, which would not need to use properties of infinite sets, would be much less problematical than the principles used by Frege to deduce, for example, the properties of numbers in his logical calculus. In this way we could be confident that, for example, number theory is consistent. The question as to whether this can be achieved is called *Hilbert's Question* which we formulate as follows.

David Hilbert, 1862–1943, was the world's leading mathematician in the years from 1900 to the 1920s. (Photo © Science Photo Library)

Don't worry if you are not familiar with the symbolism. It will be explained later.

Hilbert's Question

Can the consistency of number theory be proved using only non-dubious principles of finitary reasoning?

We shall give an answer to this question as well as Leibniz's Question in *Unit 8*, when we have assembled all that we need in order to discuss Gödel's Incompleteness Theorems, which, as we mentioned earlier, form the main goal of this half of the course.

Although at first sight there is no direct connection between Leibniz's Question and Hilbert's Question, it will be noted that Hilbert's idea exploits the fact that the rules of Frege's logical calculus are mechanical and make no use of the meaning of the formulas. Here 'mechanical' means the same as 'algorithmic', so the theory of algorithms underlies the whole enterprise and that is why we turn to it first in the remainder of this unit and in *Units 2* and *3*.

In this unit we shall explain algorithms in terms of a particular computer, called an *Unlimited Register Machine*, which can carry out the instructions in an algorithm. We shall investigate programs for the machine and look at the functions of natural numbers which they compute. In *Units 2* and *3* we shall obtain an alternative description of the functions which this machine can compute, a description which doesn't involve the machine or programs for it. We shall also obtain results of great importance about the limitations on what can be computed by this, or indeed any other, machine.

1 ALGORITHMS AND MACHINES

In this section we discuss briefly what is meant by the term *algorithm* before going on to introduce *Unlimited Register Machines* and the programs that they use to perform calculations.

1.1 What is an algorithm?

We cannot begin to answer Leibniz's Question until we have a clear idea of what an algorithm is. We are all familiar with examples of algorithms since we meet them early on in mathematics. For example, there is the method of 'long multiplication' for doing a calculation such as 874×345. Anyone who knows the method can do this calculation in a routine or mechanical way. All we have to do is to follow a procedure consisting of a sequence of simple instructions. There is no need for any deep thought. In *Unit 1* of the Number Theory part of M381 you met the Euclidean Algorithm for determining the greatest common divisor of two positive integers. Again this involves a finite sequence of prescribed steps to obtain the answer. This is typical of all algorithms. Roughly speaking, an *algorithm* is a process which produces the answer after a finite sequence of simple steps which are carried out according to specific instructions.

Since an algorithm is just the sort of process we can envisage being done by a machine, our analysis of an algorithm will be in terms of a very simple kind of calculating machine (or computer). We first describe the machine. Our claim will then be that algorithms correspond exactly with the processes that can be carried out by this machine. This claim has two parts: firstly that any process carried out by the machine is algorithmic, and secondly that any algorithmic process can be carried out by the machine we describe. The truth of the first claim will follow almost immediately from the simple nature of the machine. The second claim is altogether deeper and justifying it will involve a good deal of work.

This ambitious claim is central to much of this half of the course.

This approach to analysing algorithms was initiated by Turing (Alan Turing, 1912–1954) in around 1936. The idealized machines which Turing described are now known as *Turing machines*. For technical reasons we have chosen not to base our study of algorithms on Turing machines. Instead we introduce an alternative idealized machine which is a little easier to work with. However, because of their historical importance, we give a brief description of Turing machines in the Appendix to this unit.

Turing machines remain of theoretical importance in some areas of computer science.

1.2 Unlimited Register Machines

Unlimited Register Machines, which henceforth, and somewhat inelegantly, we abbreviate as URMs, were introduced in a paper by J.C. Shepherdson and H.E. Sturgis published in 1963. URMs are often regarded as easier to understand than Turing machines because they resemble more closely the working of digital computers. We emphasize, however, that no previous knowledge of computing is needed to understand them, but we do allow ourselves the occasional remark which we hope will be illuminating for those with some knowledge in this area.

Our treatment of URMs follows very closely that in the highly recommended book *Computability* by Nigel Cutland (Cambridge University Press, 1980).

We begin with a description of URMs. It may not be apparent at first what is going on, but the examples which follow the description should make things clearer.

An *Unlimited Register Machine* has a number of locations where it can store numbers. These locations are called *registers*. Any given URM program will make use of only a specific finite number of these registers, but we do not fix any upper bound for the number of registers that a URM program can use. The numbers that can be stored in a URM register are *natural numbers*: that is, numbers taken from the set

$$\mathbb{N} = \{0, 1, 2, 3, \ldots\} = \{n \in \mathbb{Z} : n \geq 0\}$$

where $\mathbb{Z}$ denotes the set of all integers.

Some books, including the Number Theory half of this course, use the term 'natural number' to mean 'positive integer'. Most logic books use the term to mean 'non-negative integer', as we do.

We use the upper-case letters $R_1, R_2, R_3, \ldots$ to refer to the registers and the corresponding lower-case letters $r_1, r_2, r_3, \ldots$ to indicate the numbers stored in those registers. So we have the following picture

R_1	R_2	R_3		R_k
r_1	r_2	r_3	$\cdots$	r_k

of a URM with k registers.

You should note that we are not concerned with the physical implementation of a URM, only with its basic structure. The registers could be electronic storage devices, or they could be squares on a blackboard, or boxes containing pebbles.

The URM manipulates the numbers stored in its registers according to a program. A *URM program* is a finite list of *basic instructions*. The instructions are written in an order and numbered 1, 2, 3, ... We begin by specifying these basic instructions. In the box below we give the name of each type of instruction, the standard notation used for the instruction and a description of what the effect of carrying out the instruction is on the numbers stored in the registers.

Definition 1.1 Basic instructions for a URM program

Name	Notation	Effect
Zero	$Z(n)$	Replace the number in R_n by 0.
Successor	$S(n)$	Add 1 to the number in R_n.
Copy	$C(m, n)$	Replace the number in R_n by the number in R_m.
Jump	$J(m, n, q)$	If the numbers in R_m and R_n are equal go to instruction number q, otherwise go to the next instruction.

When a URM executes a program it always starts by carrying out the first instruction of the program. When it has carried out one instruction it moves to the next instruction, unless required otherwise by a Jump instruction. We adopt the convention that a URM program stops when there is no next instruction to carry out. This can happen in two ways. If the program carries out the last instruction and this does not involve a jump to an earlier instruction, then the computation will stop. Also, if the URM carries out a Jump instruction which involves jumping to a non-existent instruction, then again this leads to the computation stopping. These points will become clearer when we give examples below.

Shortly we shall introduce *flow diagrams* to describe the implementation of a URM program. These will take the form of blocks representing the basic instructions linked by arrows to indicate the order in which the basic instructions are implemented. The blocks for the basic instructions are shown in the box below.

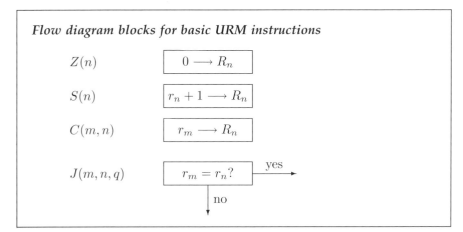

Flow diagram blocks for basic URM instructions

$Z(n)$ $\quad$ $0 \longrightarrow R_n$

$S(n)$ $\quad$ $r_n + 1 \longrightarrow R_n$

$C(m,n)$ $\quad$ $r_m \longrightarrow R_n$

$J(m,n,q)$ $\quad$ $r_m = r_n?$ $\quad$ yes

$\quad$ no

Those familiar with computing will notice that the basic instructions of the URM programming language resemble those of standard programming languages. In such languages the instructions which we have written as $Z(n)$, $S(n)$, $C(m,n)$ and $J(m,n,q)$ would appear in such forms as $R_n := 0$, $R_n := R_n + 1$, $R_n := R_m$ and

$\quad$ if $R_m \neq R_n$ do next else do q.

At first sight our programming language for URMs looks very weak and you might expect that we cannot use it to carry out very complicated calculations. A key intention of this unit is to show that this first impression is completely wrong. As we shall see, the URM programming language is extremely powerful.

There is another respect in which URMs may seem limited. They handle only natural numbers, whereas computers can handle positive and negative decimals as well as doing word-processing, producing graphic displays and so on. However, whatever may appear on the computer monitor, the processing that goes on inside the computer is all in terms of 0s and 1s and, in this sense, URMs are no more restrictive than computers.

There is one respect in which URMs go beyond computers. We have placed no restriction on the size of the numbers that can be stored in a URM register. That is why they are called *Unlimited* Register Machines. In any physical machine there will be some upper bound to the size of numbers that can be stored and numerical overflow is a real problem in practical computing. However, allowing arbitrarily large numbers in registers is only a mild idealization.

Example 1.1

Here is our first example of a URM program.

> 1 $J(2,3,5)$
> 2 $S(1)$
> 3 $S(3)$
> 4 $J(1,1,1)$

The following points about this program should be observed.

(a) This program uses only the registers R_1, R_2 and R_3.

(b) If $r_2 = r_3$, the first instruction involves a jump to instruction 5, which does not exist, so in this situation the computation halts.

(c) The final instruction involves jumping back to the first instruction if $r_1 = r_1$ and so, as this is always true, the effect of this instruction is the unconditional jump 'go to instruction number 1'.

'Go to' instructions can always be formulated in this way.

(d) The overall effect of the first and last instructions is that the program includes a 'do while $r_2 \neq r_3$' loop.

'Do while' loops can always be formulated in this way.

Representing each instruction by the corresponding block, this program has the following *flow diagram*.

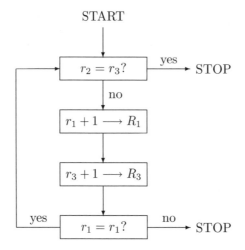

As the answer to '$r_1 = r_1$?' is always yes, we can omit the lowest box above and instead write the flow diagram more concisely as follows.

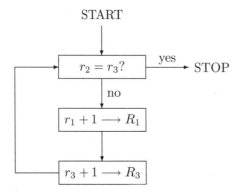

We give now a sample computation using this program. In order to show how the computation proceeds, we need to keep track of both the numbers in the registers at each stage of the computation and the number of the instruction about to be carried out. We show this in a table called the *trace table* of the computation. Let us perform the computation using the above program when the numbers initially in the registers are as shown.

7	2	0

Note that we have omitted the labels R_1, R_2, R_3 for the registers as we shall often do henceforth.

The corresponding trace table is as follows.

Instruction	R_1	R_2	R_3
1	7	2	0
2	7	2	0
3	8	2	0
4	8	2	1
1	8	2	1
2	8	2	1
3	9	2	1
4	9	2	2
1	9	2	2
STOP	9	2	2

Notice that in each row of the table we give the number of the instruction *which is about to be carried out* together with the numbers in each of the registers which the program uses *before* that instruction is carried out. For example, the first row indicates that the program is to carry out the first instruction with the numbers in the registers being 7, 2, 0. Since the number in register R_2 is not equal to the number in register R_3, no jump is carried out and the program moves to the second instruction without any change being made to the numbers in the registers. The second instruction involves adding 1 to the number in register R_1 and then going on to the next instruction. So the third row of the trace table shows that the program is about to carry out instruction 3 with the numbers in the registers now being 8, 2, 0. In the last-but-one row of the table, the program is about to carry out the first instruction with 9, 2, 2 in the registers. Since now the number in R_2 is equal to the number in R_3, the jump to instruction 5 is carried out, and, because there is no instruction 5, this means that the computation comes to an end, as we have indicated by putting 'STOP' in the bottom row. ♦

Problem 1.1

(a) For each of the following initial contents of the registers R_1, R_2 and R_3, carry out the computation using the URM program of Example 1.1.

(i) 7 0 0   (ii) 7 1 0   (iii) 7 3 0

(b) What do you think would happen if the initial contents of the registers R_1, R_2 and R_3 were as follows, where n, m are any natural numbers?

n m 0

Thus from the solution to Problem 1.1(b) it seems that, in some sense, the program of Example 1.1 carries out addition. In what sense will be made precise in the next section, but we note that this provides the first evidence that our programming language is more powerful than it might at first have seemed.

Problem 1.2 _____

Draw a flow diagram for the following URM program.

We shall discuss what this program does in the next section.

$$
\begin{array}{ll}
1 & J(1,3,12) \\
2 & J(2,3,11) \\
3 & C(1,3) \\
4 & S(5) \\
5 & J(2,5,11) \\
6 & Z(4) \\
7 & S(3) \\
8 & S(4) \\
9 & J(1,4,4) \\
10 & J(1,1,7) \\
11 & C(3,1)
\end{array}
$$

2 URM-COMPUTABLE FUNCTIONS

In Section 1 you were introduced to URMs and saw a URM program for adding two natural numbers. In Subsection 2.1 you will see URM programs for a variety of mathematical functions — examples of the so-called *URM-computable functions*. In Subsection 2.2 we shall start a systematic investigation of which functions are URM-computable. You will be introduced to three basic URM-computable functions and to one of the ways in which other URM-computable functions can be constructed from these.

2.1 Definition and examples

Recall that a *function* associates with each element of one set (the *domain* of the function) an element of a second set (the *codomain*). We write $f : A \longrightarrow B$ to indicate that f is a function with domain A and codomain B, and $x \longmapsto f(x)$ to show that f associates $f(x)$ in B with x in A; $f(x)$ is said to be the *value of f at x*.

In the Mathematical Logic half of the course the functions we are concerned with all have as their codomain the set $\mathbb{N}$ of natural numbers. Their domains will be either $\mathbb{N}$ or the set of all ordered pairs (n_1, n_2) of natural numbers or, more generally, the set of all *ordered k-tuples* $(n_1, n_2, \ldots, n_k)$ of natural numbers. Just as we use $\mathbb{R}^k$ to denote the set of points of k-dimensional space whose elements are given as ordered k-tuples of real numbers, so we use $\mathbb{N}^k$ for the set of ordered k-tuples of natural numbers: that is,

Recall that in the Mathematical Logic units $\mathbb{N}$ is the set $\{0, 1, 2, 3, \ldots\}$.

$$\mathbb{N}^k = \{(n_1, n_2, \ldots, n_k) : n_j \in \mathbb{N}, \ j = 1, 2, \ldots, k\}.$$

Thus we shall be concerned with functions with domain $\mathbb{N}^k$ and codomain $\mathbb{N}$. We refer to a function $f : \mathbb{N}^k \longrightarrow \mathbb{N}$ as a *function of k variables*. For example, addition can be regarded as the function add: $\mathbb{N}^2 \longrightarrow \mathbb{N}$ of two variables given by

When $k = 2$ we shall often prefer to use the notation (n, m) rather than (n_1, n_2) for a general element of $\mathbb{N}^2$.

$$\mathrm{add}(n_1, n_2) = n_1 + n_2.$$

Now suppose that we wish to compute values of a function of k variables. Then we must be able to input ordered k-tuples of numbers. Our *input convention* will be that to input the ordered k-tuple $(n_1, n_2, \ldots, n_k)$ we begin the computation with the number n_1 in register R_1, n_2 in $R_2, \ldots, n_k$ in R_k and 0 in all the other registers which the program uses:

R_1	R_2		R_k	R_{k+1}	R_{k+2}	
n_1	n_2	$\cdots$	n_k	0	0	$\cdots$

So, for example, the trace table in Example 1.1 can be regarded as the trace table of a computation with input $(7, 2)$.

As we are concerned with the computation of values of functions with codomain $\mathbb{N}$ we need an output which is a single number. By convention we take the *output* to be the number that is in register R_1 when the computation halts. There is nothing very special about using this particular register for the output. It is just a matter of convenience. If the result of a given program is that the answer which we want to be the output ends up in some other register, say R_s, then by adding to the end of the program the Copy instruction $C(s, 1)$ we can ensure that we obtain the desired output in register R_1 as our convention demands.

Now we are able to explain what we mean by saying that a particular URM program computes a particular function.

Definition 2.1 URM-computability

We say that the URM program P *computes* the function $f : \mathbb{N}^k \longrightarrow \mathbb{N}$ if, for all ordered k-tuples $(n_1, n_2, \ldots, n_k) \in \mathbb{N}^k$, the computation of a URM using the program P with input $(n_1, n_2, \ldots, n_k)$ produces the output $f(n_1, n_2, \ldots, n_k)$. The function $f : \mathbb{N}^k \longrightarrow \mathbb{N}$ is said to be *URM-computable* if there is a URM program which computes it.

Example 2.1

The addition function add: $\mathbb{N}^2 \longrightarrow \mathbb{N}$ is URM-computable.

We saw in Problem 1.1 that the computation using the URM program

> 1 $J(2, 3, 5)$
> 2 $S(1)$
> 3 $S(3)$
> 4 $J(1, 1, 1)$

This is the program in Example 1.1.

with the initial contents of registers R_1, R_2 and R_3 respectively as n, m and 0, halts with $n + m$ in register R_1. As the program doesn't refer to any of the registers R_s for $s > 3$, the same would happen if the initial contents of these registers were also 0, as for R_3. Thus, following our input and output conventions for computing a function of two variables, this program computes the function add: $\mathbb{N}^2 \longrightarrow \mathbb{N}$.   ♦

Note that the computation halts with the answer in R_1, as required by the convention for the output, but possibly also with non-zero numbers in other registers. This is typical for most URM programs computing functions and will need to be borne in mind later in the unit, when we seek to combine URM programs.

In this course we are much more interested in general theoretical results about URM-computable functions than in devising programs to compute particular functions or in working out which function a particular URM program computes. However, we include the following examples and problems to help you build up a feel for the power of the URM programming language.

If you feel that you need extra practice with URMs, there are further problems in the additional exercises at the end of the unit.

Having shown that addition is URM-computable, we next consider whether multiplication is a URM-computable function. We can multiply two natural numbers n and m by performing repeated additions $n + n$, $n + n + n$ and so on until we have a sum with m terms. To achieve a URM program for multiplication all we need to do is to use the idea of the program of Example 1.1 to do additions, and to keep track of how many times we do this.

Example 2.2

As we shall see below, the following URM program computes the multiplication function.

$$
\begin{array}{ll}
1 & J(1,3,12) \\
2 & J(2,3,11) \\
3 & C(1,3) \\
4 & S(5) \\
5 & J(2,5,11) \\
6 & Z(4) \\
7 & S(3) \\
8 & S(4) \\
9 & J(1,4,4) \\
10 & J(1,1,7) \\
11 & C(3,1)
\end{array}
$$

♦

To help understand how the program of Example 2.2 works, we shall try it out with some example inputs. Here is the trace table of the computation using this program with input $(2,3)$.

The flow diagram for the program is given in Solution 1.2.

Instruction	R_1	R_2	R_3	R_4	R_5
1	2	3	0	0	0
2	2	3	0	0	0
3	2	3	0	0	0
4	2	3	2	0	0
5	2	3	2	0	1
6	2	3	2	0	1
7	2	3	2	0	1
8	2	3	3	0	1
9	2	3	3	1	1
10	2	3	3	1	1
7	2	3	3	1	1
8	2	3	4	1	1
9	2	3	4	2	1
4	2	3	4	2	1
5	2	3	4	2	2
6	2	3	4	2	2
7	2	3	4	0	2
8	2	3	5	0	2
9	2	3	5	1	2
10	2	3	5	1	2
7	2	3	5	1	2
8	2	3	6	1	2
9	2	3	6	2	2
4	2	3	6	2	2
5	2	3	6	2	3
11	2	3	6	2	3
STOP	6	3	6	2	3

The output of this computation is 6, the final value in register R_1.

Problem 2.1

Give the trace tables for the computations using the URM program of Example 2.2 with the following inputs.

(a) $(3,0)$ (b) $(3,1)$ (c) $(3,2)$

After doing Problem 2.1 you should have a feel for the way the program of Example 2.2 works. The computation begins with n in R_1, m in R_2 and 0 in registers R_3, R_4 and R_5. The jump of instruction 1 occurs only if $n = 0$, and it is to a non-existent instruction so that the computation stops and the output is the content of R_1, namely 0. The jump of instruction 2 occurs only if $n \neq 0$ and $m = 0$; and, because of Copy instruction 11, the output is the number initially in R_3, namely 0. Now suppose that $n \neq 0$ and $m \neq 0$. The first steps (Copy instruction 3 and Successor instruction 4) put n into R_3 and 1 into R_5. Thus the contents of the registers at this stage are

n	m	n	0	1

If $m \neq 1$, a loop made up of instructions 7, 8, 9 and 10 adds n to the content of R_3. When this has been achieved, instruction 9 gives a jump to the Successor instruction 4; 1 is added to the content of R_5 and the contents of the registers are

n	m	$2n$	n	2

The effect of instruction 5 is to check whether or not the content of register R_5 is equal to m. If not there is another loop to add n to R_3. Notice that the content of register R_4 is used as a counter in this loop, so the Zero instruction 6 sets it to zero before the loop begins. After $m - 1$ applications of this process, the jump to instruction 4 and the implementation of instruction 4, the contents of the registers are

n	m	nm	n	m

The Jump instruction 5 leads to an application of Copy instruction 11 and the computation halts with output the final content of register R_1, namely nm. Thus the program of Example 2.2 computes the multiplication function $\text{mult} : \mathbb{N}^2 \longrightarrow \mathbb{N}$ given by $\text{mult}(n, m) = nm$.

Example 2.3

We consider how to devise a URM program which computes the *minimum* function $\min : \mathbb{N}^2 \longrightarrow \mathbb{N}$ defined by

$$\min(n, m) = \begin{cases} m, & \text{if } m \leq n, \\ n, & \text{if } n < m. \end{cases}$$

We shall later also use the *maximum* function $\max : \mathbb{N}^2 \longrightarrow \mathbb{N}$ defined by

$$\max(n, m) = \begin{cases} n, & \text{if } m \leq n, \\ m, & \text{if } n < m. \end{cases}$$

Suppose we start a computation with n in register R_1 and m in R_2. If we start with 0 in register R_3 and use the Successor instruction to increment it by 1 at a time, at each stage comparing it with the numbers in R_1 and R_2, then the first match tells us which is the smaller of n and m. Then by jumping to the appropriate instruction we can arrange for the output to be n or m, whichever is the smaller. This suggests that we might try a program along the following lines.

    1    $S(3)$
    2    $J(1, 3, 6)$
    3    $J(2, 3, 5)$
    4    $J(1, 1, 1)$
    5    $C(2, 1)$

If you work out the output of this program with inputs such as $(3, 5)$ or $(7, 2)$ you will see that it outputs the smaller of the two input numbers in each case.

However, the program is not quite correct. Since it begins by incrementing the value in R_3 by 1, if either $n = 0$ or $m = 0$ the instructions 2 and 3 will fail to detect this and we shall not obtain the correct output. This defect can be remedied by changing the order of the first three instructions. We thus end up with the following program.

Indeed if both $n = 0$ and $m = 0$ the computation will never stop. The possibility that the computation using a URM program might not stop for certain inputs is one that will concern us in *Unit 3*.

1   $J(1, 3, 6)$
2   $J(2, 3, 5)$
3   $S(3)$
4   $J(1, 1, 1)$
5   $C(2, 1)$

It has the following flow diagram.

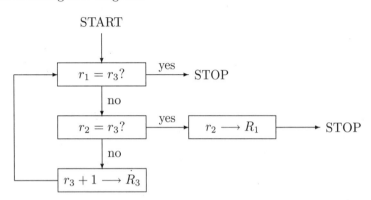

You may wish to check that this program computes the function min.   ◆

Problem 2.2

Consider the following URM program.

1   $J(1, 4, 9)$
2   $S(3)$
3   $J(1, 3, 7)$
4   $S(2)$
5   $S(3)$
6   $J(1, 1, 3)$
7   $C(2, 1)$

(a) Give the trace table of the computation with this program for the following single-number inputs. Specify the output in each case.

(i) 0   (ii) 1   (iii) 4

(b) Which function of one variable is computed by this program? *Hint*: It might help to draw a flow diagram.

You might have noticed that, in the program of Problem 2.2, the first instruction involves a jump to a non-existent instruction, namely instruction 9. The program would have had the same effect if the first instruction had been of the form

1   $J(1, 4, k)$

for any $k \geq 8$, given that the program only has 7 instructions. As you will see later, for some purposes it can be helpful if any jumps to non-existent instructions are to an instruction number one greater than the final actual instruction of the program, 8 in this case; but in general any number $k \geq 8$ will do.

Problem 2.3 _____

Consider the following URM program.

$$
\begin{array}{ll}
1 & J(2,3,9) \\
2 & J(1,3,9) \\
3 & S(3) \\
4 & S(4) \\
5 & J(2,4,7) \\
6 & J(1,1,2) \\
7 & Z(4) \\
8 & J(1,1,2) \\
9 & C(4,1)
\end{array}
$$

(a) Give the trace table of the computation with this program for the following inputs of pairs of numbers. Specify the output in each case.

   (i) $(7,3)$ (ii) $(4,2)$ (iii) $(5,0)$

(b) Which function of two variables is computed by this program?

Problem 2.4 _____

By giving suitable URM programs show that the following functions are URM-computable.

(a) $n \longmapsto 3$

(b) $n \longmapsto \begin{cases} 0, & \text{if } n = 0, \\ 1, & \text{if } n \neq 0. \end{cases}$

(c) $(n, m) \longmapsto \begin{cases} 0, & \text{if } n = m, \\ 1, & \text{if } n \neq m. \end{cases}$

(d) $(n, m) \longmapsto 3n + 5m$

(e) $(n, m) \longmapsto |n - m|$

You may have noticed that, although we can easily see from a URM program which registers it uses, we cannot tell whether the program has been designed to compute a function of one variable or a function of two variables and so on. In fact the same program could compute functions of *any* finite number of variables depending on the input. For example, a URM program which computes the function of two variables $(n, m) \longmapsto 3n + 5m$, as in Problem 2.4(d), will also compute the function of one variable $n \longmapsto 3n$.

The idea that a URM program can compute a function of any number of variables (depending on the input) will prove important in *Unit 3.*

2.2 Building new URM-computable functions

URMs provide a good theoretical model for studying algorithms but the approach of showing that large numbers of functions are URM-computable by inventing more and more URM programs has several disadvantages. As the functions get more complicated, the work of inventing correct programs for computing them becomes more difficult. Even with the simple example of the minimum function, our first attempt at a program had a defect in it. As programs get longer and longer it is harder to see what they are doing and harder to check that they do compute correctly the values of the function in which we are interested. Compare, for example, how easy it is to understand the formulas describing the functions in Problem 2.4 with the difficulty of understanding the programs in Problems 2.2 and 2.3.

Therefore we adopt a different approach which builds on our ability to describe functions by mathematical formulas. We show first that certain functions given by very simple formulas are URM-computable, and then we show that the processes for building up formulas correspond to operations on functions which yield URM-computable functions from URM-computable functions. For example, we shall be able to prove that if the functions $f : \mathbb{N}^2 \longrightarrow \mathbb{N}$ and $g : \mathbb{N}^2 \longrightarrow \mathbb{N}$ are URM-computable, then so also is the function $f + g : \mathbb{N}^2 \longrightarrow \mathbb{N}$ given by $(n, m) \longmapsto f(n, m) + g(n, m)$. Once we have proved this it will follow immediately from what we already know that such functions as $(n, m) \longmapsto nm + n + m$ and $(n, m) \longmapsto 3nm + 5(n + m)$ are URM-computable.

There is a similarity here with the way we approach the theory of continuous functions in the area of mathematics known as real analysis. In real analysis we give a technical definition of a *continuous* function. When we are faced with a complicated function such as $x \longmapsto x^3 \sin(2x^2 + 3x + 5)$, instead of proving directly from the definition that this function is continuous we use the fact that certain basic functions, for example $x \longmapsto x$ and $x \longmapsto \sin x$, are continuous and various rules, for example the Sum, Product and Composition Rules, for generating new continuous functions from functions that we already know to be continuous.

> If you are not familiar with this theory you can ignore this paragraph without missing anything essential.

Thus our approach will be to begin by showing that certain very simple functions are URM-computable. Then we show that certain operations yield URM-computable functions when applied to URM-computable functions. Another way to put this is that the set of URM-computable functions is *closed* under certain operations. This will give a powerful array of machinery for proving that functions are URM-computable.

The simple functions that we begin with can all be computed by URM programs consisting of a single instruction. It is clear that corresponding to the Zero instruction $Z(1)$ we get the *zero function* which we write as zero: $n \longmapsto 0$ and corresponding to the Successor instruction $S(1)$ we get the *successor function* succ: $n \longmapsto n + 1$. Next we think about the functions that we can obtain using Copy instructions. A Copy instruction of the form $C(m, n)$ with $n \neq 1$ does not affect the number in register R_1 (the output register) and so we shall only consider Copy instructions of the form $C(m, 1)$. The effect of the one-instruction program

> 1 $C(m, 1)$

is to replace the number originally in register R_1 with the number originally in register R_m. The function which this program computes depends on how many variables we allow for the input. If we have at least m variables in the input, the program produces as output the mth value in the ordered k-tuple of numbers making up the input: that is, if we regard the program as computing a function of k variables, where $m \leq k$, that function is

> If we have fewer than m variables in the input, the program produces the output 0.

$$(n_1, n_2, \ldots, n_k) \longmapsto n_m.$$

By analogy with the geometrical case where the function $f : \mathbb{R}^2 \longrightarrow \mathbb{R}$ given by $f(x, y) = x$ maps each point of the plane onto its projection on the x-axis, we call functions of this type *projection functions*. For historical reasons they are denoted by U. Thus for $m \leq k$, U_m^k is the projection function defined by

> Note that U_1^1 is the *identity function* id: $n \longmapsto n$, which may thus be regarded as a projection function.

$$U_m^k(n_1, n_2, \ldots, n_k) = n_m.$$

Theorem 2.1

The following functions are URM-computable.

(a) The zero function

$$\text{zero}: \mathbb{N} \longrightarrow \mathbb{N}$$
$$n \longmapsto 0$$

(b) The successor function

$$\text{succ}: \mathbb{N} \longrightarrow \mathbb{N}$$
$$n \longmapsto n + 1$$

(c) The projection functions

$$U_m^k: \mathbb{N}^k \longrightarrow \mathbb{N}$$
$$(n_1, n_2, \ldots, n_k) \longmapsto n_m$$

for all positive integers m, k with $m \leq k$.

Proof

We need only observe that these functions are computed by the following URM programs.

(a)   1 $Z(1)$

(b)   1 $S(1)$

(c)   1 $C(m, 1)$ ∎

Problem 2.5

Investigate the functions of *one* variable computed by one-instruction programs other than the programs specified in the proof of Theorem 2.1. *Hint*: Recall that, in the Mathematical Logic part of the course, a function of one variable is defined to have as domain the *whole* of $\mathbb{N}$.

We discuss now processes that we can use to build new URM-computable functions from those we already have. We begin with the straightforward operation of forming the *composite* of two functions of one variable. If $f : \mathbb{N} \longrightarrow \mathbb{N}$ and $g : \mathbb{N} \longrightarrow \mathbb{N}$ are functions, then the composite function $f \circ g$ is defined by

$$(f \circ g)(n) = f(g(n)).$$

To compute $(f \circ g)(n)$, we first compute the value of the function g with input n, and then compute the value of f with input $g(n)$. Thus we might expect that if we have URM programs for computing the functions g and f then, by putting the instructions of a program which computes f after the instructions of a program which computes g, we should obtain a URM program for computing the composite function $f \circ g$. This is more or less correct, but we have to watch out for a few technical difficulties.

To make things specific we work with the following two URM programs.

We shall look at other ways of building new URM-computable functions in Section 3.

Program P

    1   $J(1, 4, 10)$
    2   $C(1, 4)$
    3   $S(2)$
    4   $J(1, 2, 10)$
    5   $Z(3)$
    6   $S(3)$
    7   $S(4)$
    8   $J(1, 3, 3)$
    9   $J(1, 1, 6)$
   10   $C(4, 1)$

This program computes the function $f : n \longmapsto n^2$.

Program Q

    1   $C(1, 3)$
    2   $J(2, 3, 10)$
    3   $S(2)$
    4   $S(1)$
    5   $S(1)$
    6   $J(1, 1, 2)$

This program computes the function $g : n \longmapsto 3n$.

If we put the instructions of the program P immediately after the instructions of the program Q we obtain the following program.

    1   $C(1, 3)$
    2   $J(2, 3, 10)$
    3   $S(2)$
    4   $S(1)$
    5   $S(1)$
    6   $J(1, 1, 2)$
    7   $J(1, 4, 10)$
    8   $C(1, 4)$
    9   $S(2)$
   10   $J(1, 2, 10)$
   11   $Z(3)$
   12   $S(3)$
   13   $S(4)$
   14   $J(1, 3, 3)$
   15   $J(1, 1, 6)$
   16   $C(4, 1)$

Does this program compute the composite function $f \circ g : n \longmapsto (3n)^2$? In fact it is easily seen that several things have gone wrong.

First, when the instructions of program P are placed after the instructions of program Q their numbers are increased by 6. So, for example, the instruction $J(1, 4, 10)$ which is the first instruction of P is now instruction 7 and $S(4)$ which was instruction 7 of P is instruction 13 in the new program. However, we have not changed the corresponding numbers in the Jump instructions of P. For example, in P the instruction $J(1, 1, 6)$ is an unconditional jump back to instruction 6 of P which, as we have seen, is instruction 12 of the new program. Thus when we put programs together in this way, we need to adjust the instruction numbers which occur in the Jump instructions of the second program.

Another problem is that we want the combined program, as soon as it has finished using program Q to compute $g(n)$, to go to the first instruction of program P to begin computing $f(g(n))$. However, the program Q stops when it carries out the instruction $J(2, 3, 10)$ involving a jump to instruction 10 of the combined programs which is not the first instruction of P. We can get over this difficulty by changing instruction 2 of Q to

$J(2, 3, 7)$. This is still a jump to a non-existent instruction of Q and hence does not change the effect of Q but, with this change, when the computation using the instructions of Q stops, the combined program immediately begins to execute the instructions of P starting with the first instruction of P.

This leads to a general recipe for combining programs.

This illustrates why it can be helpful in the design of a program if any jumps to non-existent instructions are to an instruction number one greater than the final actual instruction of the program.

Definition 2.2 Concatenation

Suppose that Q and P are URM programs and that Q has s instructions and P has r instructions. The *concatenation* of Q and P, which is written as $Q * P$, is the program obtained in the following way.

1　Replace every Jump instruction of P of the form $J(m, n, q)$ by $J(m, n, q + s)$.

2　Replace every Jump instruction of Q of the form $J(m, n, q)$ where $q > s$ by $J(m, n, s + 1)$.

3　Write the instructions of the amended program P under the instructions of the amended program Q, renumbering them $s + 1$ to $s + r$.

Read $Q * P$ as the instructions of Q first, followed by those of P, with suitable adjustments to some of the instruction numbers. This is at variance with the notation $f \circ g$ for composition of functions, where it is g that is applied first. When combining programs, which are just lists, it is much more natural to read the combination as 'left first'.

Example 2.4

For the programs Q and P considered above, $Q * P$ is the program

1	$C(1, 3)$
2	$J(2, 3, 7)$
3	$S(2)$
4	$S(1)$
5	$S(1)$
6	$J(1, 1, 2)$
7	$J(1, 4, 16)$
8	$C(1, 4)$
9	$S(2)$
10	$J(1, 2, 16)$
11	$Z(3)$
12	$S(3)$
13	$S(4)$
14	$J(1, 3, 9)$
15	$J(1, 1, 12)$
16	$C(4, 1)$

♦

Does the revised program of Example 2.4 compute the composite function $f \circ g : n \longmapsto (3n)^2$? Unfortunately there is one more small technical difficulty. When we say that the program P computes the function $f : n \longmapsto n^2$ we mean that if it starts its computation with the register contents as

n	0	0	0	$\cdots$

then when the computation stops the number n^2 is in the register R_1. Thus for this program to produce the output $(3n)^2$ it must begin its computation with the configuration

$3n$	0	0	0	$\cdots$

However, the effect of the program Q which computes the function $g : n \longmapsto 3n$ is that its computation halts with the register contents as

$3n$	n	n	0	$\cdots$

Thus in the computation using the concatenated program $Q * P$, the part of the computation using the instructions of P does not start with the correct configuration of its registers. Therefore $Q * P$ may not compute the function $f \circ g : n \longmapsto (3n)^2$ as desired.

In fact, $Q * P$ computes $n \longmapsto 6n^2$.

Fortunately this difficulty is easily overcome. All we have to do is to insert between the instructions of Q and those of P the instructions $Z(2)$ and $Z(3)$ which put 0 in registers R_2 and R_3 or, in other words, which clear those registers. These new instructions will be numbered 7 and 8, so the instructions of P will need to be renumbered 9 to 18, and each Jump instruction of P of the form $J(m, n, q)$ must be replaced by $J(m, n, q + 8)$.

It is convenient to use the abbreviation $Z(a, b)$, where a and b are positive integers with $a \le b$, to stand for the URM program

$$
\begin{array}{ll}
1 & Z(a) \\
2 & Z(a+1) \\
\vdots & \vdots \\
b - a + 1 & Z(b)
\end{array}
$$

which clears all the registers $R_a, R_{a+1}, \ldots, R_b$. It is convenient also to adopt the convention that if $a > b$ then $Z(a, b)$ is the empty list of instructions; that is, $Z(a, b)$ does not contain any instructions at all.

Example 2.5

With the notation we have been using in this subsection, a program which computes the composite function $f \circ g : n \longmapsto (3n)^2$ is the program $(Q * Z(2, 3)) * P$: that is, the program

$$
\begin{array}{ll}
1 & C(1,3) \\
2 & J(2,3,7) \\
3 & S(2) \\
4 & S(1) \\
5 & S(1) \\
6 & J(1,1,2) \\
7 & Z(2) \\
8 & Z(3) \\
9 & J(1,4,18) \\
10 & C(1,4) \\
11 & S(2) \\
12 & J(1,2,18) \\
13 & Z(3) \\
14 & S(3) \\
15 & S(4) \\
16 & J(1,3,11) \\
17 & J(1,1,14) \\
18 & C(4,1)
\end{array}
$$
♦

It is now very easy to generalize Example 2.5 to show that the set of URM-computable functions of one variable is closed under composition. To do this we shall need to replace $Z(2, 3)$ in the example by the program which clears all the registers used by Q other than R_1. For this purpose it is useful to introduce notation for the largest register number used by a URM program P. We use $\rho(P)$ for this number. Thus in any program P no instruction refers to any register beyond R_u, where $u = \rho(P)$. In other words, all instructions in P have the forms $Z(n)$, $S(n)$, $C(m, n)$ or $J(m, n, q)$ with $m, n \le \rho(P)$.

ρ is the Greek letter 'rho'.

Theorem 2.2 Closure under composition

If $f : \mathbb{N} \longrightarrow \mathbb{N}$ and $g : \mathbb{N} \longrightarrow \mathbb{N}$ are URM-computable functions of one variable then the composite function $f \circ g : \mathbb{N} \longrightarrow \mathbb{N}$, $n \longmapsto f(g(n))$, is URM-computable.

Proof

Suppose that $f : \mathbb{N} \longrightarrow \mathbb{N}$ and $g : \mathbb{N} \longrightarrow \mathbb{N}$ are URM-computable functions of one variable. Let P be a URM program which computes f and let Q be a URM program which computes g. Put $u = \rho(Q)$. After the program Q has computed the value of $g(n)$ we need to clear all the registers used by Q other than R_1, where the value of $g(n)$ is stored. This is accomplished by the program $Z(2, u)$. Therefore, in the light of the definition of concatenation, the program

$$(Q * Z(2, u)) * P$$

computes the composite function $f \circ g$. Hence this function is URM-computable. ∎

Problem 2.6 _____

Write down a URM program to compute the function $g \circ f$ where g and f are the functions $n \longmapsto 3n$ and $n \longmapsto n^2$ respectively.

Problem 2.7 _____

Suppose that R, Q and P are URM programs. Is the concatenated program $R * (Q * P)$ the same as the program $(R * Q) * P$?

_____

We shall occasionally need to concatenate more than two programs in the next section, so that the positive answer to Problem 2.7 will be helpful, as it stops us from worrying about brackets!

3 MORE TECHNIQUES

In Section 2 you saw how new URM-computable functions of one variable can be constructed from others by composition. In this section we shall introduce two other ways of constructing new URM-computable functions, namely *substitution* and *primitive recursion*, both of which can be used for functions of more than one variable.

3.1 Substitution

Our aim in this subsection is to generalize Theorem 2.2 about closure of URM-computable functions under composition to cover cases of functions of more than one variable.

We begin with the simplest extension of Theorem 2.2. Suppose $g_1 : \mathbb{N} \longrightarrow \mathbb{N}$, $g_2 : \mathbb{N} \longrightarrow \mathbb{N}$ and $f : \mathbb{N}^2 \longrightarrow \mathbb{N}$ are all URM-computable functions and let $h : \mathbb{N} \longrightarrow \mathbb{N}$ be defined by

$$h(n) = f(g_1(n), g_2(n)).$$

Since there are URM programs which compute the values of g_1, g_2 and f, it seems plausible that we can use these to construct a URM program which computes the values of h. The proof of this is similar to that of Theorem 2.2. We have just to be more careful about the technicalities.

A program which computes the value of $h(n)$ needs to compute successively the values of $g_1(n)$, $g_2(n)$ and $f(g_1(n), g_2(n))$. When it comes to compute $g_2(n)$, it will need to be supplied with the value of n. The program which computes $g_1(n)$ will start with n in register R_1 and will end with $g_1(n)$ in this register. So unless we take care to store n in some register not used in the calculation, it will be lost and not available for the calculation of $g_2(n)$.

Likewise we need to store $g_1(n)$ so that it can be used for the calculation of $f(g_1(n), g_2(n))$. We can achieve this by copying these values to registers not used in the calculations. And, as in the proof of Theorem 2.2, after computing $g_1(n)$ we must clear those registers used in the computation of $g_1(n)$, other than R_1 which holds the value of $g_1(n)$; similarly, after computing $g_2(n)$ we must clear those registers other than R_1 used in the computation of $g_2(n)$. This strategy is implemented in the proof of the following theorem.

Theorem 3.1

Suppose that $g_1 : \mathbb{N} \longrightarrow \mathbb{N}$, $g_2 : \mathbb{N} \longrightarrow \mathbb{N}$ and $f : \mathbb{N}^2 \longrightarrow \mathbb{N}$ are URM-computable functions. Then the function $h : \mathbb{N} \longrightarrow \mathbb{N}$

$$n \longmapsto f(g_1(n), g_2(n))$$

is also URM-computable.

Proof

Let P, Q_1 and Q_2 be URM programs which compute the functions f, g_1 and g_2 respectively. Let u be the maximum of $\rho(P)$, $\rho(Q_1)$ and $\rho(Q_2)$. So the registers R_{u+1} and R_{u+2} are not used by any of the programs P, Q_1 and Q_2 and thus can be used to store copies of n and $g_1(n)$.

We give below a program which will compute the values of h. We have listed the instructions on the left of the page. These are to be regarded as concatenated, so that instruction numbers need to be adjusted accordingly. This includes adjusting any Jump instruction of Q_1 to a non-existent instruction so that it jumps to the first instruction immediately following Q_1; and similarly for Q_2.

To illustrate how this program works we have added an explanation in the second column, and in the third column we have indicated the register contents *after* the relevant part of the program has been executed. Where we are uninterested in the precise content of a register, we have written the content as $\diamond$. The $\diamond$ symbol will usually represent different numbers in different registers.

		R_1	R_2	R_3	$\cdots$	R_u	R_{u+1}	R_{u+2}
		n	0	0	$\cdots$	0	0	0
$C(1, u+1)$	Store n in register R_{u+1}	n	0	0	$\cdots$	0	n	0
Q_1	Compute $g_1(n)$	$g_1(n)$	$\diamond$	$\diamond$	$\cdots$	$\diamond$	n	0
$C(1, u+2)$	Store $g_1(n)$ in register R_{u+2}	$g_1(n)$	$\diamond$	$\diamond$	$\cdots$	$\diamond$	n	$g_1(n)$
$C(u+1, 1)$	Return n to R_1	n	$\diamond$	$\diamond$	$\cdots$	$\diamond$	n	$g_1(n)$
$Z(2, u)$	Clear registers $R_2, R_3, \ldots, R_u$	n	0	0	$\cdots$	0	n	$g_1(n)$
Q_2	Compute $g_2(n)$	$g_2(n)$	$\diamond$	$\diamond$	$\cdots$	$\diamond$	n	$g_1(n)$
$C(1, 2)$	Put $g_2(n)$ in R_2	$g_2(n)$	$g_2(n)$	$\diamond$	$\cdots$	$\diamond$	n	$g_1(n)$
$C(u+2, 1)$	Return $g_1(n)$ to R_1	$g_1(n)$	$g_2(n)$	$\diamond$	$\cdots$	$\diamond$	n	$g_1(n)$
$Z(3, u)$	Clear registers $R_3, \ldots, R_u$	$g_1(n)$	$g_2(n)$	0	$\cdots$	0	n	$g_1(n)$
P	Compute $f(g_1(n), g_2(n))$	$h(n)$	$\diamond$	$\diamond$	$\cdots$	$\diamond$	n	$g_1(n)$

It should be evident that the program above does indeed compute the function h. ∎

Problem 3.1

Let $f : \mathbb{N}^2 \longrightarrow \mathbb{N}$, $g_1 : \mathbb{N} \longrightarrow \mathbb{N}$ and $g_2 : \mathbb{N} \longrightarrow \mathbb{N}$ be the functions $(n, m) \longmapsto n + m$, $n \longmapsto n^2$ and $n \longmapsto 3n$ respectively and let $h : \mathbb{N} \longrightarrow \mathbb{N}$ be given by $h(n) = f(g_1(n), g_2(n))$ (that is, $h(n) = n^2 + 3n$).

Write down a URM program to compute the function h.

A URM program to compute f was given in Example 1.1 and URM programs to compute g_1 and g_2 are given as programs P and Q in Subsection 2.2.

We can think of the function h considered in Problem 3.1 in the following way. We take the function f given by the formula

$$f(n, m) = n + m$$

and we substitute $g_1(n)$ for n and $g_2(n)$ for m. This gives the formula for h. We say therefore that h is obtained from the functions f, g_1 and g_2 by *substitution*. This can be considerably generalized by allowing f to be a function of any finite number of variables, say t variables. Then we take t functions $g_1, g_2, \ldots, g_t$ to substitute for these variables. For simplicity we shall insist that these functions are all of the same number of variables, say k, but allow k to be any positive integer. In this way we reach the following definition of functions obtained by substitution.

This is not restrictive in any way, since if g is a function of r variables we can consider it to be a function of any number of variables $q \geqslant r$ (in which the additional variables play no part). Thus if g_i is a function of k_i variables, we can simply take $k = \max\{k_i : i = 1, 2, \ldots, t\}$ and consider each g_i to be a function of k variables.

Definition 3.1 Substitution

Let $f : \mathbb{N}^t \longrightarrow \mathbb{N}$, $g_1 : \mathbb{N}^k \longrightarrow \mathbb{N}$, $g_2 : \mathbb{N}^k \longrightarrow \mathbb{N}$, ..., $g_t : \mathbb{N}^k \longrightarrow \mathbb{N}$ be functions. The function $h : \mathbb{N}^k \longrightarrow \mathbb{N}$ defined by

$$h(n_1, n_2, \ldots, n_k) =$$
$$f(g_1(n_1, n_2, \ldots, n_k), g_2(n_1, n_2, \ldots, n_k), \ldots, g_t(n_1, n_2, \ldots, n_k))$$

is said to be obtained from $f, g_1, g_2, \ldots, g_t$ by *substitution*.

Problem 3.2

Given the functions $f : (n_1, n_2, n_3) \longmapsto 2n_1 + 3n_2 + 4n_3$,
$g_1 : (n_1, n_2) \longmapsto n_1 n_2$, $g_2 : (n_1, n_2) \longmapsto |n_1 - n_2|$ and
$g_3 : (n_1, n_2) \longmapsto n_1 + n_2$, let h be obtained from f, g_1, g_2, g_3 by substitution. Calculate the values of $h(2, 4)$ and $h(5, 2)$.

The proof of Theorem 3.1 can be extended to deal with general substitutions by taking care with the technical details, namely that there are more variables and more intermediate values that need to be stored. We state the theorem but omit the proof.

Theorem 3.2 Closure under substitution

If the functions $f : \mathbb{N}^t \longrightarrow \mathbb{N}$, $g_1 : \mathbb{N}^k \longrightarrow \mathbb{N}$, $g_2 : \mathbb{N}^k \longrightarrow \mathbb{N}$, ..., $g_t : \mathbb{N}^k \longrightarrow \mathbb{N}$ are all URM-computable and $h : \mathbb{N}^k \longrightarrow \mathbb{N}$ is defined from them by substitution, then h is also URM-computable.

This theorem essentially supersedes Theorem 2.2 as substitution incorporates composition of functions, which is equivalent to substitution with $t = k = 1$.

Although Theorem 3.2 gives us a method of generating new URM-computable functions, the simple use of substitution starting from the URM-computable functions of Theorem 2.1 (that is, the functions zero, succ, U_m^k) does not get us very far. For example, Theorems 2.1 and 3.2 by themselves are not powerful enough to prove that the addition function is URM-computable. We need other methods for generating URM-computable functions. We turn to this in the next subsection.

3.2 Primitive recursion

In the previous subsection we saw how to define new functions by substitution and that this process applied to URM-computable functions always produces another URM-computable function. We want to find additional processes for constructing functions which also have this property.

A good starting point is to consider a function of one variable which looks as though it could not be obtained using substitution but which also seems to be URM-computable. We consider the *factorial* function fac : $n \longmapsto n!$. The natural way to calculate the values of this function is by successive multiplications. For example

$$10! = 10 \times 9 \times 8 \times 7 \times 6 \times 5 \times 4 \times 3 \times 2 \times 1$$

and thus to calculate 10! we would calculate successively 2×1, $3 \times (2 \times 1)$, $4 \times (3 \times (2 \times 1))$ and so on. Since we already know that URMs can carry out multiplications, it seems plausible that the factorial function is URM-computable.

This method for calculating $n!$ by successive multiplications exploits the fact that we can calculate the values of $n!$ by using the formula

$$(n+1)! = (n+1) \times n!$$

which enables us successively to calculate the values of 2!, 3!, 4!, ... from the previous value in the list. Of course a starting point is needed and we recall that by convention $0! = 1$. This method of specifying the values of a function is called *definition by primitive recursion*.

Suppose that we wish to define a function $h : \mathbb{N} \longrightarrow \mathbb{N}$ by primitive recursion. As a starting point we need to know the value of $h(0)$. Then given n and $h(n)$ we need to be able to determine the value of $h(n+1)$: that is, we need to associate with the ordered pair $(n, h(n))$ the value of $h(n+1)$. A way to achieve this is to specify a function $g : \mathbb{N}^2 \longrightarrow \mathbb{N}$ such that $g(n, h(n)) = h(n+1)$. This leads to the following definition.

Definition 3.2 *Primitive recursion for a function of one variable*

Let a be a natural number and let $g : \mathbb{N}^2 \longrightarrow \mathbb{N}$ be a function. The function $h : \mathbb{N} \longrightarrow \mathbb{N}$ is said to be defined by *primitive recursion* from the constant a and the function g if

$$h(0) = a,$$
$$h(n+1) = g(n, h(n)).$$

Example 3.1

The function fac : $\mathbb{N} \longrightarrow \mathbb{N}$ given by $\text{fac}(n) = n!$ is defined by primitive recursion from the constant 1 and the function $g : \mathbb{N}^2 \longrightarrow \mathbb{N}$ given by $g(n, m) = (n+1) \times m$, since

$$\text{fac}(0) = 1,$$
$$\text{fac}(n+1) = (n+1) \times \text{fac}(n) = g(n, \text{fac}(n)). \qquad \blacklozenge$$

Example 3.2

Let $h : \mathbb{N} \longrightarrow \mathbb{N}$ be the function defined by primitive recursion from the constant 3 and the function g given by $g(n, m) = 3n + 2m$. We calculate the value of $h(5)$.

The definition by primitive recursion tells us that

$$h(0) = 3,$$
$$h(n + 1) = g(n, h(n)) = 3n + 2h(n).$$

We use these equations to give us successively the values

$$h(0) = 3,$$
$$h(1) = 0 + 2h(0) = 6,$$
$$h(2) = 3 + 2h(1) = 15,$$
$$h(3) = 6 + 2h(2) = 36,$$
$$h(4) = 9 + 2h(3) = 81,$$
$$h(5) = 12 + 2h(4) = 174.$$

When calculating $h(1), h(2), h(3), \ldots$, it often pays off to write

$$h(1) = h(0 + 1),$$
$$h(2) = h(1 + 1),$$
$$h(3) = h(2 + 1),$$

and so on, to avoid errors when substituting for n in the equation $h(n + 1) = g(n, h(n))$.

Problem 3.3

Let $h : \mathbb{N} \longrightarrow \mathbb{N}$ be defined by primitive recursion from the constant 5 and the function $g : \mathbb{N}^2 \longrightarrow \mathbb{N}$ given by $g(n, m) = n^2 + 1 + m$. Calculate $h(6)$.

This explains how functions of one variable can be defined by primitive recursion. We can extend definition by primitive recursion to functions of more than one variable. The simplest way to do this is to regard all the variables but the last as fixed parameters whose values do not change in the primitive recursion equations. Thus to define a function $h : \mathbb{N}^{k+1} \longrightarrow \mathbb{N}$ by primitive recursion, we need to specify $h(n_1, n_2, \ldots, n_k, 0)$ and to specify $h(n_1, n_2, \ldots, n_k, n + 1)$ in terms of $n_1, n_2, \ldots, n_k, n$ and $h(n_1, n_2, \ldots, n_k, n)$. This leads to the following definition.

Definition 3.3 *Primitive recursion for a function of several variables*

Let $f : \mathbb{N}^k \longrightarrow \mathbb{N}$ and $g : \mathbb{N}^{k+2} \longrightarrow \mathbb{N}$ be functions. The function $h : \mathbb{N}^{k+1} \longrightarrow \mathbb{N}$ is said to be obtained from f and g by *primitive recursion* if

$$h(n_1, n_2, \ldots, n_k, 0) = f(n_1, n_2, \ldots, n_k),$$
$$h(n_1, n_2, \ldots, n_k, n + 1) = g(n_1, n_2, \ldots, n_k, n, h(n_1, n_2, \ldots, n_k, n)).$$

The definition of primitive recursion for a function of one variable can be considered a special case of this definition for a function of several variables, with $k = 0$ and the function f replaced by the constant a.

Example 3.3

The function $g : \mathbb{N}^3 \longrightarrow \mathbb{N}$ is defined by $g(n_1, n, m) = (n_1 + m)(n + 1)$ and the function $f : \mathbb{N} \longrightarrow \mathbb{N}$ is defined by $f(n_1) = n_1^2$. Then the function $h : \mathbb{N}^2 \longrightarrow \mathbb{N}$ obtained from f and g by primitive recursion is given by the equations

$$h(n_1, 0) = f(n_1) = n_1^2,$$
$$h(n_1, n + 1) = g(n_1, n, h(n_1, n)) = (n_1 + h(n_1, n))(n + 1).$$

We calculate the value of $h(2, 5)$ by calculating successively $h(2, 0)$, $h(2, 1), \ldots$. Thus we obtain

$$h(2, 0) = 4,$$
$$h(2, 1) = (2 + h(2, 0))(0 + 1) = 6,$$
$$h(2, 2) = (2 + h(2, 1))(1 + 1) = 16,$$
$$h(2, 3) = (2 + h(2, 2))(2 + 1) = 54,$$
$$h(2, 4) = (2 + h(2, 3))(3 + 1) = 224,$$
$$h(2, 5) = (2 + h(2, 4))(4 + 1) = 1130.$$

As earlier, when calculating $h(n_1, 1), h(n_1, 2), h(n_1, 3), \ldots$, it often pays off to write

$$h(n_1, 1) = h(n_1, 0 + 1),$$
$$h(n_1, 2) = h(n_1, 1 + 1),$$
$$h(n_1, 3) = h(n_1, 2 + 1),$$

and so on, to avoid errors when substituting for n in the equation $h(n_1, n + 1) = g(n_1, n, h(n_1, n))$.

Problem 3.4 _____

(a) Let h be the function defined in Example 3.3. Calculate the values of $h(4,3)$ and $h(3,4)$.

(b) The function $f \colon \mathbb{N}^2 \longrightarrow \mathbb{N}$ is defined by $f(n_1, n_2) = n_1 + n_2$ and the function $g \colon \mathbb{N}^4 \longrightarrow \mathbb{N}$ is defined by $g(n_1, n_2, n, m) = n + (n_2 \times m)$. Let the function $h \colon \mathbb{N}^3 \longrightarrow \mathbb{N}$ be obtained from f and g by primitive recursion. Compute the values of $h(4,3,2)$ and $h(5,1,3)$.

_____

It should be evident from Example 3.3 and Problem 3.4 that provided we have algorithms for computing the values of the functions f and g then we can derive from them an algorithm for computing the values of the function h obtained from f and g by primitive recursion. To turn this into a proof that if f and g are URM-computable then so also is h, we need only be careful about storing the values of the variables. The URM program given in the proof to follow mirrors the calculation of Example 3.3. That is, to calculate the value of $h(n_1, \ldots, n_k, n)$ we use the first recursion equation to calculate $h(n_1, \ldots, n_k, 0)$ and then we use the second recursion equation to calculate the values of $h(n_1, \ldots, n_k, 1), \ldots, h(n_1, \ldots, n_k, n)$. This involves putting the value of the recursion variable equal to 0 and successively incrementing it by 1 until it reaches the value n. In this way we are led to the proof of the following theorem.

Theorem 3.3 Closure under primitive recursion for a function of several variables

Let $f : \mathbb{N}^k \longrightarrow \mathbb{N}$ and $g : \mathbb{N}^{k+2} \longrightarrow \mathbb{N}$ be URM-computable functions. Then the function $h : \mathbb{N}^{k+1} \longrightarrow \mathbb{N}$ obtained from f and g by primitive recursion is also URM-computable.

Proof

Let A and B be URM programs which compute the functions f and g respectively. Suppose that A has r instructions and B has s instructions. Let u be the maximum of $\rho(A)$, $\rho(B)$ and $k + 2$. Thus the registers $R_{u+1}, R_{u+2}, \ldots$ are neither used by the programs A and B nor for input, and so can safely be used to store values.

We construct a program to compute the function h. If the input is $(n_1, \ldots, n_k, n)$, the program will calculate the values of $h(n_1, \ldots, n_k, i)$ for $i = 0, \ldots, n$. The value of n will be stored in register R_{u+k+1} and the current value of the recursion variable i will be stored in register R_{u+k+2}. The computation ends when the numbers stored in these registers are equal.

We give the program below as a list of instructions which includes those of the programs A and B. As in the proof of Theorem 3.1, these latter programs are to be regarded as concatenated with the other instructions, so that their instruction numbers and some of their Jump instructions will get amended accordingly. We have added some comments to indicate what each part of the program is doing and to keep track of the instruction numbers and contents of the registers. As earlier, where we are uninterested in the precise content of a register, we have written the content as $\diamond$, which will usually represent different numbers in different registers.

The initial contents of the registers are as follows.

R_1		R_k	R_{k+1}	R_{k+2}	R_{k+3}	
n_1	$\cdots$	n_k	n	0	0	$\cdots$

The first $k+1$ instructions of the program are

$$1 \quad C(1, u+1)$$
$$\vdots \quad \vdots$$
$$k+1 \quad C(k+1, u+k+1)$$

These store $n_1, \ldots, n_k, n$ in the registers R_{u+1}, ..., R_{u+k}, R_{u+k+1}. The contents of the registers after performing these instructions are

R_1		R_k	R_{k+1}	R_{k+2}		R_u	R_{u+1}		R_{u+k}	R_{u+k+1}	R_{u+k+2}
n_1	$\cdots$	n_k	n	0	$\cdots$	0	n_1	$\cdots$	n_k	n	0

The next instruction is

$$k+2 \quad Z(k+1)$$

This clears register R_{k+1}.

Next we concatenate

$$A$$

with its instructions appropriately relabelled and possibly some of its Jump instructions adjusted. This computes $f(n_1, \ldots, n_k) = h(n_1, \ldots, n_k, 0)$. The contents of the registers after performing A are

R_1	R_2		R_u	R_{u+1}		R_{u+k}	R_{u+k+1}	R_{u+k+2}
$h(n_1, \ldots, n_k, 0)$	$\diamond$	$\cdots$	$\diamond$	n_1	$\cdots$	n_k	n	0

where the $\diamond$s in registers R_2 up to R_u stand for whatever numbers end up in these registers after running the program A.

The next instruction is

$$k+r+3 \quad J(u+k+1, u+k+2, k+r+u+s+6)$$

As you will see shortly this jump, if implemented, is a jump to a non-existent instruction so that if $n = 0$ then the computation stops with the correct output $h(n_1, \ldots, n_k, 0)$.

The r instructions of A get relabelled as $k+2+j$ for $j = 1, \ldots, r$. Any Jump instruction of A to a non-existent instruction is adjusted to jump to instruction $k+r+3$.

Otherwise there are successive loops in which the content of R_{u+k+2} is incremented by 1. The instructions for this loop are instruction $k + r + 3$ followed by

$$
\begin{array}{ll}
k+r+4 & C(1, k+2) \\
k+r+5 & C(u+k+2, k+1) \\
k+r+6 & C(u+1, 1) \\
\quad\vdots & \quad\vdots \\
2k+r+5 & C(u+k, k) \\
& Z(k+3, u) \\
k+r+u+4 & S(u+k+2) \\
& B \\
k+r+u+s+5 & J(1, 1, k+r+3)
\end{array}
$$

Here we are concatenating the program B, which has s instructions. The s instructions of B get relabelled as $k + r + u + j + 4$ for $j = 1, \ldots, s$.

This ends the program, so you can now see that the earlier jump to instruction $k + r + u + s + 6$ is indeed to a non-existent instruction. The effect of this loop is as follows. Suppose a loop starts with the following register contents at instruction $k + r + 4$.

R_1	R_2		R_u	R_{u+1}		R_{u+k}	R_{u+k+1}	R_{u+k+2}
$h(n_1, \ldots, n_k, i-1)$	$\diamond$	$\cdots$	$\diamond$	n_1	$\cdots$	n_k	n	$i-1$

Then after implementation of instruction $k + r + u + 4$ the contents of the registers are

R_1		R_k	R_{k+1}	R_{k+2}	R_{k+3}		R_u	R_{u+1}		R_{u+k}	R_{u+k+1}	R_{u+k+2}
n_1	$\cdots$	n_k	$i-1$	$h(n_1, \ldots, n_k, i-1)$	0	$\cdots$	0	n_1	$\cdots$	n_k	n	i

The program B then computes

$$ g(n_1, \ldots, n_k, i-1, h(n_1, \ldots, n_k, i-1)) = h(n_1, \ldots, n_k, i) $$

and then there is an unconditional jump to instruction $k + r + 3$. The contents of the registers are then

R_1	R_2		R_u	R_{u+1}		R_{u+k}	R_{u+k+1}	R_{u+k+2}
$h(n_1, \ldots, n_k, i)$	$\diamond$	$\cdots$	$\diamond$	n_1	$\cdots$	n_k	n	i

The looping ends when the contents of R_{u+k+1} and R_{u+k+2} are the same, that is when $i = n$, in which case instruction $k + r + 3$ causes a jump to a non-existent instruction. The output is then $h(n_1, \ldots, n_k, n)$ as required.

Thus the function h is URM-computable. ∎

Example 3.4

We shall use the recipe given in the proof of Theorem 3.3 to obtain a URM program to compute the function $h : \mathbb{N}^2 \longrightarrow \mathbb{N}$ obtained by primitive recursion from the zero function $\text{zero} : \mathbb{N} \longrightarrow \mathbb{N}$, $n_1 \longmapsto 0$, and the function $g : \mathbb{N}^3 \longrightarrow \mathbb{N}$, $(n_1, n, m) \longmapsto n_1 + m$.

This recipe doesn't always give the shortest or simplest URM program for h, but it does provide a systematic method for producing a correct program.

The program A which computes the zero function is

$$ 1 \quad Z(1) $$

and a program B which computes the function g is

You should work out how the program B is obtained from the program to compute addition given in Example 1.1.

$$
\begin{array}{ll}
1 & C(3, 2) \\
2 & Z(3) \\
3 & J(2, 3, 7) \\
4 & S(1) \\
5 & S(3) \\
6 & J(1, 1, 3)
\end{array}
$$

In the notation of the proof of Theorem 3.3 we have $k = 1$, $r = 1$, $s = 6$ and $u = 3$. The required program is

1	$C(1,4)$
2	$C(2,5)$
3	$Z(2)$
4	$Z(1)$
5	$J(5,6,17)$
6	$C(1,3)$
7	$C(6,2)$
8	$C(4,1)$
9	$S(6)$
10	$C(3,2)$
11	$Z(3)$
12	$J(2,3,16)$
13	$S(1)$
14	$S(3)$
15	$J(1,1,12)$
16	$J(1,1,5)$

Note that, as $k + 3 = 4 > u$, $Z(k+3, u)$ is the empty list of instructions.

Notice that the recursion equations satisfied by h are

$$h(n_1, 0) = 0,$$
$$h(n_1, n+1) = n_1 + h(n_1, n).$$

An easy argument using Mathematical Induction shows that h is the multiplication function $\mathrm{mult} : (n_1, n) \longmapsto n_1 \times n$. ◆

Problem 3.5

Let $f : \mathbb{N}^2 \longrightarrow \mathbb{N}$ be the function computed by the URM program

1	$J(2,3,5)$
2	$S(1)$
3	$S(3)$
4	$J(1,1,1)$

Let $g : \mathbb{N}^4 \longrightarrow \mathbb{N}$ be the function computed by the URM program

1	$S(4)$
2	$Z(1)$
3	$C(4,1)$

Let the function $h : \mathbb{N}^3 \longrightarrow \mathbb{N}$ be obtained from f and g by primitive recursion.

(a) Use the recipe given in the proof of Theorem 3.3 to obtain a URM program to compute h.

(b) What in general is the value of $h(n_1, n_2, n)$?

There is an analogue to Theorem 3.3 for functions of one variable.

Theorem 3.4 Closure under primitive recursion for a function of one variable

Let $a \in \mathbb{N}$ be a constant and let $g : \mathbb{N}^2 \longrightarrow \mathbb{N}$ be a URM-computable function. Then the function $h : \mathbb{N} \longrightarrow \mathbb{N}$ obtained from a and g by primitive recursion is also URM-computable.

This theorem can be considered as a special case of Theorem 3.3 with $k = 0$ and the function f replaced by the constant a.

You are asked to prove this theorem in one of the exercises for this section.

APPENDIX: TURING MACHINES

Because of their historical importance, in this appendix we give a brief account of *Turing machines*. This material will not be assessed or examined.

These machines were devised by Alan Turing around 1936 and provided the first characterization of algorithms in terms of idealized computing machines. Turing's ideas began with a philosophical analysis of what was involved in a mechanical, or algorithmic, computation. This was before electronic digital computers had been invented, so for Turing a 'computer' meant a human calculator, not a machine. As Turing saw it, such a computer works by manipulating symbols on a piece of paper following specific mechanical rules. Although mathematical calculations are usually carried out on a two-dimensional piece of paper, to simplify things Turing imagined his machine working in one dimension, on a tape divided into squares.

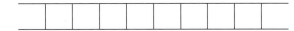

Turing envisaged that each square could either be blank or could contain a symbol taken from a specific finite list of symbols, say $s_1, s_2, \ldots, s_a$. The machine would examine just one square at a time and would carry out an action determined by the symbol in the square and the current internal state of the machine.

The *internal states* of the machine provide a device whereby the machine, which only examines one square at a time, can keep some track of the symbols in other squares. There would be a finite number of possible internal states, say $q_1, q_2, \ldots, q_b$.

The actions that the machine can carry out at any stage are of three kinds.

(a) Replace the symbol in the square by another symbol.

(b) Move to examine the square to the immediate left of the current square being looked at.

(c) Move to examine the square to the immediate right of the current square being looked at.

After carrying out an action, the machine may move into a different internal state.

The program for such a machine consists of a set of instructions specifying what action to carry out in *some* possible combinations of the internal state and the symbol in the square currently being examined. The instructions also specify the state the machine moves into after carrying out the action. The instructions thus have the form

$$q_i \quad s_j \quad A \quad q_t$$

which we interpret as 'if the machine is in the internal state q_i and the symbol in the square being examined is s_j, then the machine should carry out the action A and move into the internal state q_t'. We use s_k to indicate the action of replacing the symbol in the square being examined by the symbol s_k. The actions of moving to examine the square to the immediate left or right of the current square are indicated by L or R, respectively. We adopt the convention that the computation stops when there is no instruction specifying what should be done in the current situation.

Example

Here is an example of a Turing machine program which uses the symbols 0 and 1 only, and three internal states q_1, q_2, q_3.

$$q_1 \quad 0 \quad R \quad q_2$$
$$q_1 \quad 1 \quad 0 \quad q_1$$
$$q_2 \quad 0 \quad 1 \quad q_3$$
$$q_2 \quad 1 \quad R \quad q_2$$

Alan Turing, 1912–1954, British mathematician, logician and codebreaker, was also credited as the father of the theory of artificial intelligence. (Photo © Science Photo Library)

Notice that there are no instructions specifying what to do when the machine is in the internal state q_3 so, by the convention described above, the computation will stop whenever the machine enters this state. Notice also that, unlike the instructions in a URM program, the order in which we write down these instructions does not matter. The instruction to be carried out is determined by the current internal state of the machine and the symbol in the square currently being examined. Where the relevant instruction for this situation occurs in the list of instructions does not matter.

We illustrate how this Turing machine program works by a sample computation. We adopt the convention that the computation starts with the machine in state q_1. The square currently being examined is indicated by putting the symbol in that square in bold print. All undisplayed squares are assumed to contain the symbol 0. The current internal state of the machine is written to the left of the diagram showing the symbols in the squares.

q_1 | 0 | **1** | 1 | 1 | 0 | 1 | 1 | 0 |

q_1 | 0 | **0** | 1 | 1 | 0 | 1 | 1 | 0 |

q_2 | 0 | 0 | **1** | 1 | 0 | 1 | 1 | 0 |

q_2 | 0 | 0 | 1 | **1** | 0 | 1 | 1 | 0 |

q_2 | 0 | 0 | 1 | 1 | **0** | 1 | 1 | 0 |

q_3 | 0 | 0 | 1 | 1 | **1** | 1 | 1 | 0 |

One possible input convention for Turing machines is to represent the positive integer n by a block of n 1s, the pair (n, m) of positive integers by blocks of n 1s and m 1s separated by a 0, and so on. A similar convention can be used for the output. Using these conventions the computation above starts with input $(3, 2)$ and has as output the single number 5. Indeed it may be checked that the program above produces the output $n + m$ for the input (n, m) and thus that it is a Turing machine program for addition.   ♦

Although the program above is very simple, it is not difficult to appreciate that devising Turing machine programs for more complicated functions is rather more difficult than is the case with URMs because Turing machines store numbers a digit at a time. Indeed, the program for addition was only straightforward because of the convention we used for inputs and outputs. Imagine devising a Turing machine program for doing addition using standard decimal notation. Such a program would need to incorporate carrying rules, so that for example the initial tape configuration

| 7 | 3 | 4 | 8 | | 1 | 7 | 6 |

would produce the output configuration

| 7 | 5 | 2 | 4 | | | | |

Alternatively, think about devising a Turing machine program for multiplication using the blocks of 1s convention for inputs and outputs that we used above with the program for addition.

These considerations demonstrate why it is easier to work with URMs than with Turing machines. None the less the particularly simple nature of Turing machines gives them a theoretical importance that we mention later. Notice also that while URMs are designed to operate with *numbers*, a Turing machine works with *strings of symbols* and so we can think of Turing machines as carrying out computations with objects other than numbers.

During the Second World War, Alan Turing worked at the Government Code and Cypher School at Bletchley Park on the task of cracking enemy cyphers. One of his contributions to this work was the invention of electronic calculating devices to aid the work. After the war he moved to the University of Manchester where he was involved in building the first British general-purpose computer. In March 1952 he was charged with committing homosexual acts which were then illegal. He pleaded guilty to these charges and was put on probation subject to the condition that he undertook a course of hormone injections intended to reduce his homosexual urges. He committed suicide on 7 June 1954 by taking cyanide.

SUMMARY

To tackle the important questions of Leibniz and Hilbert, stated in the Introduction and which we shall eventually answer in *Unit 8*, a rigorous definition of what is a mechanical or algorithmic process of calculation is required. We introduced a theoretical computer called an *Unlimited Register Machine* (URM) and explained how it can be programmed to perform calculations. We claimed that algorithms correspond exactly with the processes which can be carried out by a URM. URM computations are certainly mechanical, but it will not be until *Unit 3* that we shall be able to justify the claim that every algorithmic process can be carried out by a URM.

We introduced the notion of a *URM-computable function* and gave examples of URM programs to compute the values of URM-computable functions. However, we are much more interested in general theoretical results about URM-computable functions, so we began a study of the way we can obtain new URM-computable functions from those already known to be URM-computable. The processes of substitution and primitive recursion were introduced. These processes give URM-computable functions when applied to URM-computable functions.

The notion of primitive recursion will be very important in the course. In *Unit 2* we shall study *primitive recursive functions*, which are those functions which can be obtained from certain basic functions (proved to be URM-computable in Theorem 2.1 of this unit) by using the operations of substitution and primitive recursion a finite number of times. It follows from results in this unit that every primitive recursive function is URM-computable. Many interesting examples of primitive recursive, and hence URM-computable, functions will be given in *Unit 2*.

OBJECTIVES

We list those things on which we may set assessment questions to test your understanding of this unit.

After working through the unit you should be able to:

(a) recall the definitions of the basic URM instructions and of a URM program;

(b) write down a trace table for the computation of a given URM program with a given input and state what is the output;

(c) draw a flow diagram for a given URM program;

(d) determine, in simple cases, which function of a given number of variables is computed by a given URM program;

(e) construct a URM program to compute a given function;

(f) write down the concatenation of given URM programs;

(g) calculate the values of a function defined by primitive recursion;

(h) write down a URM program to compute a function obtained by substitution or primitive recursion from given URM-computable functions.

ADDITIONAL EXERCISES

Most of these exercises provide further practice, should you feel you need it, in handling the main ideas in the unit on which you are likely to be assessed.

There are a few harder problems, labelled as such in the margin. These are harder than any of the problems you are likely to encounter in the assessment and are included solely as challenges for the interested student.

Section 1

1 Determine which of the following are URM programs. In the cases which are not URM programs, explain why.

(a) 1 $J(1,2,5)$
     2 $S(3,4)$
     3 $C(2,1)$
     4 $J(1,1,1)$

(b) 1 $C(1,1)$
     2 $C(1,3)$
     3 $C(3,1)$
     4 $C(3,1)$

(c) 1 $J(1,2,5)$
     2 $T(1,3)$
     3 $C(3,1)$
     4 $J(1,1,8)$

(d) 1 $J(1,2,5)$
     2 $S(2)$
     3 $J(1,1,1,1)$
     4 $J(1,1,8)$

2 Show that, for all positive integers m, n with $m \neq n$, there is a URM program which does not use any Copy instructions but which has the same effect as the Copy instruction $C(m,n)$.

Section 2

1   In each of the following cases, determine which function of *one* variable is computed by the given URM program.

(a) 1 $S(1)$
      2 $S(1)$
      3 $S(1)$
      4 $S(1)$
      5 $S(1)$

(b) 1 $Z(1)$
      2 $S(1)$
      3 $S(1)$

(c) 1 $J(1,2,8)$
      2 $S(2)$
      3 $S(3)$
      4 $J(1,2,8)$
      5 $S(2)$
      6 $Z(3)$
      7 $J(1,1,1)$
      8 $C(3,1)$

(d) 1 $J(1,2,7)$
      2 $S(2)$
      3 $J(1,2,7)$
      4 $S(2)$
      5 $S(3)$
      6 $J(1,1,1)$
      7 $C(3,1)$

In cases (c) and (d) you might find it helpful to do some sample computations or to draw a flow diagram.

2   Devise URM programs to compute the following functions of one variable.

(a) $n \longmapsto n+2$

(b) $n \longmapsto 5$

(c) $n \longmapsto$ the remainder when n is divided by 3

(d) $n \longmapsto \begin{cases} n-2, & \text{if } n \geq 2, \\ 0, & \text{otherwise.} \end{cases}$

3   In each of the following cases determine which function of *two* variables is computed by the given URM program.

(a) 1 $J(1,3,5)$
      2 $J(2,3,6)$
      3 $S(3)$
      4 $J(1,1,1)$
      5 $C(2,1)$

(b)   1 $J(2,3,6)$
       2 $S(4)$
       3 $S(4)$
       4 $S(3)$
       5 $J(1,1,1)$
       6 $J(1,5,10)$
       7 $J(4,5,11)$
       8 $S(5)$
       9 $J(1,1,6)$
     10 $C(4,1)$

(c) 1 $J(1,3,6)$
     2 $S(4)$
     3 $S(4)$
     4 $S(3)$
     5 $J(1,1,1)$
     6 $Z(3)$
     7 $J(2,3,13)$
     8 $S(4)$
     9 $S(4)$
    10 $S(4)$
    11 $S(3)$
    12 $J(1,1,7)$
    13 $C(4,1)$

4   Devise URM programs to compute the following functions of two variables.

(a) $(n,m) \longmapsto n + m + 2$

(b) $(n,m) \longmapsto \begin{cases} 1, & \text{if } n \leq m, \\ 0, & \text{otherwise.} \end{cases}$

5   Devise a URM program to compute the function of three variables

$$(n_1, n_2, n_3) \longmapsto n_1 + 2n_2 + 3n_3.$$

6   Prove that if a function of k variables is URM-computable then it can be computed by a URM program which does not include any Zero instructions.

7   Devise a URM program which when started with 0 in all registers finishes with 9 in register R_1. Can you find such a URM program with fewer than 9 instructions?

Harder problem

8   Investigate those functions of one variable that can be computed by URM programs which do *not* include: (a) any Jump instructions; (b) any Successor instructions. (Ideally, you should aim to characterize the set of functions which can be computed by URM programs which do not include any Jump instructions and the set of functions which can be computed by URM programs which do not include any Successor instructions.)

Harder problem

Section 3

1   Devise a URM program to compute the function $n \longmapsto (n+1)(n+2)$.

2   The function $f: \mathbb{N}^2 \longrightarrow \mathbb{N}$ is defined by $f(n_1, n_2) = n_2$ and the function $g: \mathbb{N}^4 \longrightarrow \mathbb{N}$ is defined by $g(n_1, n_2, n, m) = (n_1 \times n) + n_2 + m$. Let the function $h: \mathbb{N}^3 \longrightarrow \mathbb{N}$ be obtained from f and g by primitive recursion. Compute the values of $h(1,5,2)$ and $h(4,2,3)$.

3   Let $h: \mathbb{N}^2 \longrightarrow \mathbb{N}$ be defined by primitive recursion from the identity function $\text{id}: \mathbb{N} \longrightarrow \mathbb{N}$ and the function $g: \mathbb{N}^3 \longrightarrow \mathbb{N}$ given by

$$g(n_1, n, m) = \begin{cases} 0, & \text{if } m = 0, \\ m - 1, & \text{otherwise.} \end{cases}$$

(a) Use the recipe in the proof of Theorem 3.3 to obtain a URM program to compute h.

(b) What in general is the value of $h(n_1, n)$?

4   Prove Theorem 3.4. *Hint*: Adapt the proof of Theorem 3.3 to take account of the fact that the function f has been replaced by a constant a. Remember also that Theorem 3.4 can be considered a special case of Theorem 3.3 with $k = 0$.

5   Use your solution to Exercise 4 to obtain a URM program to compute the function $n \longmapsto 2^n$.

SOLUTIONS TO THE PROBLEMS

Solution 1.1

(a) (i)

Instruction	R_1	R_2	R_3
1	7	0	0
STOP	7	0	0

(ii)

Instruction	R_1	R_2	R_3
1	7	1	0
2	7	1	0
3	8	1	0
4	8	1	1
1	8	1	1
STOP	8	1	1

(iii)

Instruction	R_1	R_2	R_3
1	7	3	0
2	7	3	0
3	8	3	0
4	8	3	1
1	8	3	1
2	8	3	1
3	9	3	1
4	9	3	2
1	9	3	2
2	9	3	2
3	10	3	2
4	10	3	3
1	10	3	3
STOP	10	3	3

(b) In each case in part (a), when the computation comes to an end the number in register R_1 is the sum of the numbers which were initially in registers R_1 and R_2. In fact if the initial content of R_3 is zero, this is always the case.

In this part the initial contents of R_1 and R_2 are the natural numbers n and m respectively and the initial content of R_3 is zero. If $m = 0$ then the computation stops at once and the content of register R_1 is $n = n + m$. Suppose that $m \neq 0$. Then 1 is added to the content of R_1 until we have done so m times. This is done with a loop and the content of register R_3 records the number of times the loop is performed. When the content of R_3 is m (that is, equal to the content of R_2) the loop has been performed m times, the computation stops and the content of register R_1 is $n + m$.

Solution 1.2

Recall that the computation will stop either when it has carried out the last instruction and this is not a Jump, as with instruction 11, or when a Jump instruction involves jumping to a non-existent instruction, as with instruction 1.

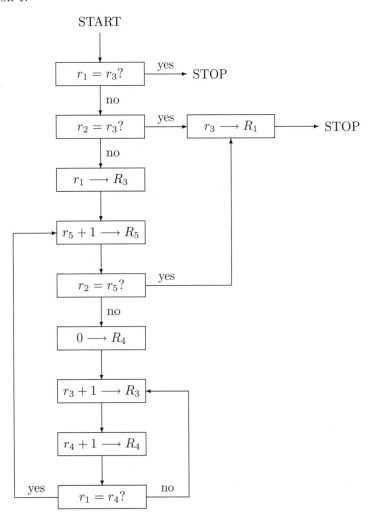

Solution 2.1

(a)

Instruction	R_1	R_2	R_3	R_4	R_5
1	3	0	0	0	0
2	3	0	0	0	0
11	3	0	0	0	0
STOP	0	0	0	0	0

(b)

Instruction	R_1	R_2	R_3	R_4	R_5
1	3	1	0	0	0
2	3	1	0	0	0
3	3	1	0	0	0
4	3	1	3	0	0
5	3	1	3	0	1
11	3	1	3	0	1
STOP	3	1	3	0	1

(c)

Instruction	R_1	R_2	R_3	R_4	R_5
1	3	2	0	0	0
2	3	2	0	0	0
3	3	2	0	0	0
4	3	2	3	0	0
5	3	2	3	0	1
6	3	2	3	0	1
7	3	2	3	0	1
8	3	2	4	0	1
9	3	2	4	1	1
10	3	2	4	1	1
7	3	2	4	1	1
8	3	2	5	1	1
9	3	2	5	2	1
10	3	2	5	2	1
7	3	2	5	2	1
8	3	2	6	2	1
9	3	2	6	3	1
4	3	2	6	3	1
5	3	2	6	3	2
11	3	2	6	3	2
STOP	6	2	6	3	2

Solution 2.2

Recall our convention that the output is the final content of R_1.

(a) (i)

Instruction	R_1	R_2	R_3	R_4
1	0	0	0	0
STOP	0	0	0	0

Output $= 0$.

(ii)

Instruction	R_1	R_2	R_3	R_4
1	1	0	0	0
2	1	0	0	0
3	1	0	1	0
7	1	0	1	0
STOP	0	0	1	0

Output $= 0$.

(iii)

Instruction	R_1	R_2	R_3	R_4
1	4	0	0	0
2	4	0	0	0
3	4	0	1	0
4	4	0	1	0
5	4	1	1	0
6	4	1	2	0
3	4	1	2	0
4	4	1	2	0
5	4	2	2	0
6	4	2	3	0
3	4	2	3	0
4	4	2	3	0
5	4	3	3	0
6	4	3	4	0
3	4	3	4	0
7	4	3	4	0
STOP	3	3	4	0

Output = 3.

(b) It is not usually possible to do just a few calculations as in part (a) and then deduce from the input–output behaviour of a URM program which function the program computes. In general you need to combine this information with some thought about how the program operates. Sometimes a flow diagram helps. The flow diagram for this program is as follows.

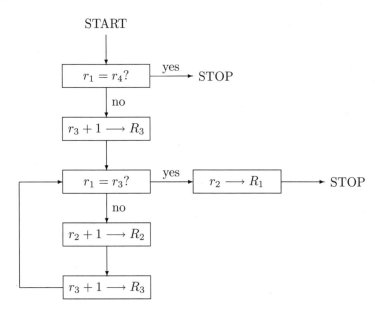

Suppose that the input is n. If $n = 0$ or $n = 1$, we have already seen that the output is 0 so we suppose that $n > 1$. The computation in part (a) with input 4 illustrates the general case. Implementation of instruction 2 puts 1 into R_3. Then there are loops through instructions 3, 4, 5, 6 in which the contents of registers R_2 and R_3 are incremented by 1. The computation stops when the content of R_3 is equal to n and the output is the content of R_2 which is $n - 1$. Thus the function f computed by this program is given by

$$f(n) = \begin{cases} 0, & \text{if } n = 0, \\ n - 1, & \text{otherwise.} \end{cases}$$

Solution 2.3

(a) (i)

Instruction	R_1	R_2	R_3	R_4
1	7	3	0	0
2	7	3	0	0
3	7	3	0	0
4	7	3	1	0
5	7	3	1	1
6	7	3	1	1
2	7	3	1	1
3	7	3	1	1
4	7	3	2	1
5	7	3	2	2
6	7	3	2	2
2	7	3	2	2
3	7	3	2	2
4	7	3	3	2
5	7	3	3	3
7	7	3	3	3
8	7	3	3	0
2	7	3	3	0
3	7	3	3	0
4	7	3	4	0
5	7	3	4	1
6	7	3	4	1
2	7	3	4	1
3	7	3	4	1
4	7	3	5	1
5	7	3	5	2
6	7	3	5	2
2	7	3	5	2
3	7	3	5	2
4	7	3	6	2
5	7	3	6	3
7	7	3	6	3
8	7	3	6	0
2	7	3	6	0
3	7	3	6	0
4	7	3	7	0
5	7	3	7	1
6	7	3	7	1
2	7	3	7	1
9	7	3	7	1
STOP	1	3	7	1

Output = 1.

(ii)

Instruction	R_1	R_2	R_3	R_4
1	4	2	0	0
2	4	2	0	0
3	4	2	0	0
4	4	2	1	0
5	4	2	1	1
6	4	2	1	1
2	4	2	1	1
3	4	2	1	1
4	4	2	2	1
5	4	2	2	2
7	4	2	2	2
8	4	2	2	0
2	4	2	2	0
3	4	2	2	0
4	4	2	3	0
5	4	2	3	1
6	4	2	3	1
2	4	2	3	1
3	4	2	3	1
4	4	2	4	1
5	4	2	4	2
7	4	2	4	2
8	4	2	4	0
2	4	2	4	0
9	4	2	4	0
STOP	0	2	4	0

Output = 0.

(iii)

Instruction	R_1	R_2	R_3	R_4
1	5	0	0	0
9	5	0	0	0
STOP	0	0	0	0

Output = 0.

(b) Here it is rather harder to see what is going on and you are encouraged to draw a flow diagram if you find this helpful.

Consider an input (n, m). If $m = 0$ the effect of the first instruction is to give output 0. Suppose $m \neq 0$. In the computation there are loops in which the contents of registers R_3 and R_4 are incremented by 1 until the content of register R_3 is equal to n. In these loops, each time the content of register R_4 reaches m it is set to zero, so we can see that if n is a multiple of m the computation stops with zero in register R_4, which is the output. Now suppose that $n = qm + r$ where $0 < r < m$. After $n - r$ loops as described above the contents of the registers will be

n	m	$n - r$	0

Then there will be r loops through instructions 2 to 6 (with no jump at instruction 5) to produce the register contents

n	m	n	r

After implementation of instructions 2 and 9, the computation stops with output the number in R_4, which is r. Thus the function f computed by this program is

$$f(n,m) = \begin{cases} 0, & \text{if } m = 0, \\ \text{the remainder when} & \\ n \text{ is divided by } m, & \text{if } m \neq 0. \end{cases}$$

Solution 2.4

(a) Here we want to ensure that, whatever the input, the output is always 3. All we need to do therefore is ensure that the computation stops with 3 in the first register. This can easily be achieved by the following program.

$$\begin{array}{ll} 1 & Z(1) \\ 2 & S(1) \\ 3 & S(1) \\ 4 & S(1) \end{array}$$

(b) First we must decide whether or not the input is 0. Then we have to ensure that when the input is 0 we jump to a part of the program that produces output 0 and that otherwise we get output 1. This is achieved by the following program.

$$\begin{array}{ll} 1 & J(1,2,5) \\ 2 & Z(1) \\ 3 & S(1) \\ 4 & J(1,1,6) \\ 5 & Z(1) \end{array}$$

(c) We can use the same program as in part (b). With input (n,m) the effect of the first instruction is now to determine whether or not $n = m$.

(d) The following program is suitable.

$$\begin{array}{ll} 1 & J(1,4,7) \\ 2 & S(3) \\ 3 & S(3) \\ 4 & S(3) \\ 5 & S(4) \\ 6 & J(1,1,1) \\ 7 & Z(4) \\ 8 & J(2,4,16) \\ 9 & S(3) \\ 10 & S(3) \\ 11 & S(3) \\ 12 & S(3) \\ 13 & S(3) \\ 14 & S(4) \\ 15 & J(1,1,8) \\ 16 & C(3,1) \end{array}$$

Suppose the input is (n,m). We accumulate the answer in register R_3. The first loop from instruction 1 to instruction 6 adds 3 to this register. We use the register R_4 to count the number of times we do this and we jump to the next part of the program when the number in R_4 is n. Instruction 7 puts 0 into R_4 so that this register can again be used as a counter in the next part of the computation. The second loop from instructions 8 to 15 adds 5 to R_3 m times. Thus we end up with $3n + 5m$ in R_3. The final instruction then copies this number to R_1 so that it becomes the output.

(e) The following program is suitable.

$$
\begin{array}{ll}
1 & C(1,3) \\
2 & C(2,4) \\
3 & J(1,4,9) \\
4 & J(2,3,9) \\
5 & S(3) \\
6 & S(4) \\
7 & S(5) \\
8 & J(1,1,3) \\
9 & C(5,1)
\end{array}
$$

Suppose that the input is (n, m). The program begins by copying n, m to the registers R_3, R_4 respectively. The loop made up of instructions 3 to 8 has the effect of adding 1 to each of these registers and also to R_5. So after the computation has been round this loop k times the contents of the registers are

n	m	$n+k$	$m+k$	k

If $m \le n$, the Jump instruction 3 is carried out when $n = m + k$; and if $m > n$, the Jump instruction 4 is carried out when $m = n + k$. In each case the output is k. So when $m \le n$ the output is $n - m$ and when $m > n$ the output is $m - n$. Thus in both cases the output is $|n - m|$ as required.

Solution 2.5

We consider the different forms the other one-instruction programs can take.

- By changing 1 to n, where $n > 1$, in each of the programs in the proof of Theorem 2.1 we obtain the following three programs.

(a) 1 $Z(n)$

(b) 1 $S(n)$

(c) 1 $C(m, n)$

Programs of these forms do not alter the number in the first register, so they all compute the identity function, that is, the function $U_1^1 : n \longmapsto n$.

- The only other one-instruction programs involve the Jump instruction. There are two possibilities.

(a) 1 $J(m, n, q)$ where $q > 1$.

Programs of these forms, whether the numbers in R_m and R_n are equal or not, stop without altering the number in the first register, so they also compute the identity function U_1^1.

(b) 1 $J(m, n, 1)$

If $r_m = r_n$ initially, then the computation does not halt but keeps carrying out the Jump instruction. So if $m = n = 1$, we have $r_m = r_n = r_1$ and the computation does not halt for any input (which of course goes into register 1). If $m = 1$ and $n \ne 1$, the computation does not halt for input 0 (i.e. $r_1 = 0$). This follows because, for all $n \ne 1$, r_n is also 0 (by the input convention, since we are considering only functions of *one* variable) and we thus have $r_m = r_n = 0$. The computation would halt for any other input, without altering the number in the first register, but the failure to halt with input 0 means that the program does not compute a function (since, as you will recall, the functions of one variable we

> Remember that you are asked to consider only functions of *one* variable.

are considering have as domain the *whole* of $\mathbb{N}$). Similarly, if $n = 1$ and $m \neq 1$, the computation does not halt for input 0. And if neither m nor n is 1, the computation does not halt for any input, as we have $r_m = r_n = 0$ for *any* input. In all cases, there is at least one input for which the computation does not halt, so that for no combination of values for m and n does the program compute a function of one variable.

If we allowed the domain of a function to be subset of $\mathbb{N}$ rather than $\mathbb{N}$ itself, the program in the case $m = 1$ and $n \neq 1$ would compute a function with domain $\{n \in \mathbb{N} : n > 0\}$. In *Unit 3* we discuss how to describe 'functions' computed by programs where the computation does not stop for certain inputs.

Solution 2.6

We use the notation already established. The program P to compute a function f uses registers R_1, R_2, R_3 and R_4, so a program which computes $g \circ f$ is $(P * Z(2, 4)) * Q$: that is,

1	$J(1, 4, 10)$
2	$C(1, 4)$
3	$S(2)$
4	$J(1, 2, 10)$
5	$Z(3)$
6	$S(3)$
7	$S(4)$
8	$J(1, 3, 3)$
9	$J(1, 1, 6)$
10	$C(4, 1)$
11	$Z(2)$
12	$Z(3)$
13	$Z(4)$
14	$C(1, 3)$
15	$J(2, 3, 23)$
16	$S(2)$
17	$S(1)$
18	$S(1)$
19	$J(1, 1, 15)$

Solution 2.7

The answer is yes!

The number, type and order of the instructions of both programs are easily seen to be the same. But we need to check that the amendments to the Jump instructions in each program have the same results.

Suppose that R and Q have t and s instructions respectively.

Let us look first at what happens to the Jump instructions of P. When forming $(Q * P)$, every Jump instruction of P of the form $J(m, n, q)$ is replaced by $J(m, n, q + s)$, so when forming $R * (Q * P)$ the instruction is replaced by $J(m, n, q + s + t)$. On the other hand, as $(R * Q)$ has $s + t$ instructions, when we form $(R * Q) * P$ every Jump instruction of P of the form $J(m, n, q)$ is replaced by $J(m, n, q + s + t)$. Thus the Jump instructions of P have been replaced by the same instructions in each of $R * (Q * P)$ and $(R * Q) * P$.

Next look at the Jump instructions of Q. When forming $(Q * P)$, every Jump instruction of Q of the form $J(m, n, q)$ where $q > s$ is replaced by $J(m, n, s + 1)$, so when forming $R * (Q * P)$ the instruction is replaced by $J(m, n, s + 1 + t)$. On the other hand, when forming $(R * Q)$, the instruction is replaced by $J(m, n, q + t)$. As $q > s$ and $(R * Q)$ has $s + t$ instructions, when we form $(R * Q) * P$ we must account for the fact that $q + t > s + t$ and replace the instruction by $J(m, n, s + t + 1)$, which agrees with the corresponding instruction in $R * (Q * P)$.

We can show similarly that Jump instructions of Q of the form $J(m, n, q)$ where $q \leq s$ and Jump instructions of R are amended in the same way in each of $R * (Q * P)$ and $(R * Q) * P$.

Solution 3.1

We follow the proof of Theorem 3.1 to construct the following URM program. The URM programs to compute f, g_1 and g_2 use the registers R_1, R_2, R_3 and R_4 only. Thus, in the notation of Theorem 3.1, we have $u = 4$.

1	$C(1,5)$
2	$J(1,4,11)$
3	$C(1,4)$
4	$S(2)$
5	$J(1,2,11)$
6	$Z(3)$
7	$S(3)$
8	$S(4)$
9	$J(1,3,4)$
10	$J(1,1,7)$
11	$C(4,1)$
12	$C(1,6)$
13	$C(5,1)$
14	$Z(2)$
15	$Z(3)$
16	$Z(4)$
17	$C(1,3)$
18	$J(2,3,23)$
19	$S(2)$
20	$S(1)$
21	$S(1)$
22	$J(1,1,18)$
23	$C(1,2)$
24	$C(6,1)$
25	$Z(3)$
26	$Z(4)$
27	$J(2,3,31)$
28	$S(1)$
29	$S(3)$
30	$J(1,1,27)$

This program is not the shortest URM program to compute h. The construction in the proof of Theorem 3.1 is designed to deal with the most general case. For example, the programs to compute g_2 and f do not use register R_4 so instructions 16 and 26 could be omitted.

Notice that changes to the Jump instructions have been made in accordance with the rules for concatenation given in Definition 2.2.

Solution 3.2

In general

$$h(n_1, n_2) = f(g_1(n_1, n_2),\ g_2(n_1, n_2),\ g_3(n_1, n_2)).$$

Hence

$$h(2, 4) = f(g_1(2, 4),\ g_2(2, 4),\ g_3(2, 4)) = f(8, 2, 6) = 46,$$

$$h(5, 2) = f(g_1(5, 2),\ g_2(5, 2),\ g_3(5, 2)) = f(10, 3, 7) = 57.$$

Solution 3.3

Since $h(0) = 5$ and $h(n + 1) = n^2 + 1 + h(n)$,

$h(1) = 0 + 1 + h(0) = 6,$
$h(2) = 1 + 1 + h(1) = 8,$
$h(3) = 4 + 1 + h(2) = 13,$
$h(4) = 9 + 1 + h(3) = 23,$
$h(5) = 16 + 1 + h(4) = 40,$
$h(6) = 25 + 1 + h(5) = 66.$

Solution 3.4

(a) We have

$h(4, 0) = 4^2 = 16,$
$h(4, 1) = (4 + h(4, 0))(0 + 1) = 20,$
$h(4, 2) = (4 + h(4, 1))(1 + 1) = 48,$
$h(4, 3) = (4 + h(4, 2))(2 + 1) = 156,$

and

$h(3, 0) = 3^2 = 9,$
$h(3, 1) = (3 + h(3, 0))(0 + 1) = 12,$
$h(3, 2) = (3 + h(3, 1))(1 + 1) = 30,$
$h(3, 3) = (3 + h(3, 2))(2 + 1) = 99,$
$h(3, 4) = (3 + h(3, 3))(3 + 1) = 408.$

(b) The function h is given by the equations

$h(n_1, n_2, 0) = f(n_1, n_2) = n_1 + n_2,$
$h(n_1, n_2, n + 1) = g(n_1, n_2, n, h(n_1, n_2, n)) = n + (n_2 \times h(n_1, n_2, n)).$

We then have

$h(4, 3, 0) = 4 + 3 = 7,$
$h(4, 3, 1) = 0 + (3 \times h(4, 3, 0)) = 21,$
$h(4, 3, 2) = 1 + (3 \times h(4, 3, 1)) = 64,$

and

$h(5, 1, 0) = 5 + 1 = 6,$
$h(5, 1, 1) = 0 + (1 \times h(5, 1, 0)) = 6,$
$h(5, 1, 2) = 1 + (1 \times h(5, 1, 1)) = 7,$
$h(5, 1, 3) = 2 + (1 \times h(5, 1, 2)) = 9.$

Solution 3.5

(a) In the notation of Theorem 3.3 we have $k = 2$, $r = 4$, $s = 3$ and $u = 4$.
   The required URM program is

$$
\begin{array}{rl}
1 & C(1,5) \\
2 & C(2,6) \\
3 & C(3,7) \\
4 & Z(3) \\
5 & J(2,3,9) \\
6 & S(1) \\
7 & S(3) \\
8 & J(1,1,5) \\
9 & J(7,8,19) \\
10 & C(1,4) \\
11 & C(8,3) \\
12 & C(5,1) \\
13 & C(6,2) \\
14 & S(8) \\
15 & S(4) \\
16 & Z(1) \\
17 & C(4,1) \\
18 & J(1,1,9)
\end{array}
$$

(b) The recursion equations are

$$f(n_1, n_2) = n_1 + n_2,$$
$$g(n_1, n_2, n, m) = m + 1.$$

The program computing the function f is the one discussed in Example 1.1.

   Thus $h(n_1, n_2, 0) = n_1 + n_2$, $h(n_1, n_2, 1) = (n_1 + n_2) + 1$,
   $h(n_1, n_2, 2) = (n_1 + n_2 + 1) + 1 = n_1 + n_2 + 2$ and in general

$$h(n_1, n_2, n) = (n_1 + n_2) + n.$$

SOLUTIONS TO ADDITIONAL EXERCISES

Section 1

1 (a) This is not a URM program as $S(3,4)$ is not a valid URM instruction.

(b) This is a valid URM program. The $C(1,1)$ instruction may be thought pointless, but it is an allowable instruction! Likewise the second $C(3,1)$ instruction is pointless but allowed!

(c) This is not a URM program as $T(1,3)$ is not a valid URM instruction.

(d) This is not a URM program as $J(1,1,1,1)$ is not a valid URM instruction.

2 The program

$$
\begin{array}{ll}
1 & Z(n) \\
2 & J(m,n,5) \\
3 & S(n) \\
4 & J(1,1,2)
\end{array}
$$

has the effect of replacing the number in register R_n by the number in register R_m. Thus it has the same effect as the Copy instruction $C(m,n)$.

Section 2

1 (a) The effect of the program is to add 5 to the number initially in register R_1. Hence this program computes the function $n \longmapsto n+5$.

(b) The effect of this program is to replace the number initially in register R_1 by the number 2. So it computes the constant function $n \longmapsto 2$.

(c) In this program, the content of register R_2 is incremented successively by 1 until $r_1 = r_2$. Let n be the input number.

If n is even, then after $\frac{1}{2}n$ loops through instructions 1 to 7, the contents of the registers are

n	n	0

and the computation stops after implementation of instructions 1 and 8; the output is the number in register R_3, which is 0.

Now suppose that n is odd. After $\frac{1}{2}(n-1)$ loops through instructions 1 to 7, the contents of the registers are

n	$n-1$	0

The effect of instructions 2 and 3 is to change the contents of the registers to

n	n	1

The computation stops after implementation of instructions 4 and 8 and the output is the number in register R_3, which is 1.

Thus this program computes the function $f : \mathbb{N} \longrightarrow \mathbb{N}$ given by

$$f(n) = \begin{cases} 0, & \text{if } n \text{ is even,} \\ 1, & \text{if } n \text{ is odd.} \end{cases}$$

(d) Let n be the input number. If n is even then after $\frac{1}{2}n$ loops through instructions 1 to 6, the contents of the registers are

n	n	$\frac{1}{2}n$

and the computation stops after implementation of instructions 1 and 7; the output is the number in register R_3, which is $\frac{1}{2}n$.

Now suppose that n is odd. After $\frac{1}{2}(n-1)$ loops through instructions 1 to 6, the contents of the registers are

n	$n-1$	$\frac{1}{2}(n-1)$

Then the number in register R_2 is incremented by 1 (instruction 2) and the computation stops after implementation of instructions 3 and 7; the output is the number in register R_3, which is $\frac{1}{2}(n-1)$.

Thus this program computes the function $f : \mathbb{N} \longrightarrow \mathbb{N}$ given by

$$f(n) = \begin{cases} \frac{1}{2}n, & \text{if } n \text{ is even,} \\ \frac{1}{2}(n-1), & \text{if } n \text{ is odd.} \end{cases}$$

2   There are lots of programs that compute the given functions. Our answers for parts (a), (b) and (c) are based on the programs in the solutions to the corresponding parts of the preceding exercise, with appropriate amendments. The program in part (d) is based on that of Problem 2.2.

(a) 1 $S(1)$
    2 $S(1)$

(b) 1 $Z(1)$
    2 $S(1)$
    3 $S(1)$
    4 $S(1)$
    5 $S(1)$
    6 $S(1)$

(c) 1 $J(1,2,11)$
    2 $S(2)$
    3 $S(3)$
    4 $J(1,2,11)$
    5 $S(2)$
    6 $S(3)$
    7 $J(1,2,11)$
    8 $S(2)$
    9 $Z(3)$
   10 $J(1,1,1)$
   11 $C(3,1)$

(d) 1 $J(1,4,9)$
    2 $S(3)$
    3 $J(1,3,9)$
    4 $S(3)$
    5 $J(1,3,9)$
    6 $S(3)$
    7 $S(2)$
    8 $J(1,1,5)$
    9 $C(2,1)$

3 (a) This program is similar to that of Example 2.3 which computes the minimum of n and m. The difference is that it produces output m when the minimum program produces output n and vice versa. Thus the program computes the maximum of n and m: that is, the function $(n, m) \longmapsto \max(n, m)$.

(b) Suppose the input is (n, m). The effect of the loop consisting of instructions 1 to 5 is to put $2m$ in register R_4. The rest of the program has the same structure as that in part (a) but applied to the numbers in R_1 and R_4. So this program computes the function $(n, m) \longmapsto \max(n, 2m)$.

(c) This program is similar that given in Solution 2.4(d). It computes the function $(n, m) \longmapsto 2n + 3m$.

4 Again there are lots of possible answers. We outline briefly the approach which led to the programs we have given.

(a) We take the standard program for addition (Example 1.1) and follow it by instructions to add 2 to the output. Thus we have the following program.

$$
\begin{array}{ll}
1 & J(2,3,5) \\
2 & S(1) \\
3 & S(3) \\
4 & J(1,1,1) \\
5 & S(1) \\
6 & S(1)
\end{array}
$$

(b) Here we use the idea of the program of Example 2.3, which computes $(n, m) \longmapsto \min(n, m)$, and adjust it to give the output 1 instead of n when $n \leq m$ and the output 0 instead of m when $n > m$. Thus we have the following program.

$$
\begin{array}{ll}
1 & J(1,3,5) \\
2 & J(2,3,8) \\
3 & S(3) \\
4 & J(1,1,1) \\
5 & Z(1) \\
6 & S(1) \\
7 & J(1,1,9) \\
8 & Z(1)
\end{array}
$$

5 We shall adapt the method used in Solution 2.4(d), exploiting the fact that register R_1 starts off with content n_1, and then adding twice the content of R_2 followed by three times the content of R_3. Our solution is as follows.

$$
\begin{array}{ll}
1 & J(2,4,6) \\
2 & S(1) \\
3 & S(1) \\
4 & S(4) \\
5 & J(1,1,1) \\
6 & Z(4) \\
7 & J(3,4,13) \\
8 & S(1) \\
9 & S(1) \\
10 & S(1) \\
11 & S(4) \\
12 & J(1,1,7)
\end{array}
$$

6   Suppose that $f : \mathbb{N}^k \longrightarrow \mathbb{N}$ is computed by the URM program P. Put $u = \max(k, \rho(P))$. Then the register R_{u+1} is not used in the computation of the function f using the program P. Hence throughout this computation the number in this register is 0. Thus the instruction $C(u + 1, n)$ has the same effect as $Z(n)$. So if we obtain the URM program P' from P by replacing each instruction of the form $Z(n)$ by the instruction $C(u + 1, n)$, then P' is a URM program with no Zero instructions which also computes the function f.

Note that it is a consequence of Additional Exercise 2 for Section 1 that if a function is URM-computable then it can be computed by a URM program which does not include any Copy instructions. But one cannot dispense with both Copy and Zero instructions, as you might like to explain!

7   One obvious URM program with 9 instructions consists of the instruction $S(1)$ repeated 9 times. There are several programs that use fewer instructions, such as the following.

      1   $C(2,3)$
      2   $C(1,2)$
      3   $S(1)$
      4   $S(1)$
      5   $S(1)$
      6   $J(3,4,1)$

8   (a) Suppose a URM program P contains no Jump instructions. Then any computation using this program just carries out each instruction in turn and then stops. So each instruction is executed just once. Thus the number in any register can be incremented by 1 a fixed finite number of times or can be replaced by zero using a Zero instruction. Moreover numbers can be copied from one register to another. It follows that, with input n, the number in a given register R_k at the end of the computation must have either the form c_k or the form $n + c_k$ where c_k is a natural number which depends on the program P but not on the input n. Thus, writing c in place of c_1, the function computed by P is either a constant function $n \longmapsto c$ or a function of the form $n \longmapsto n + c$.

Conversely it is easily seen that all functions of these forms are computable by URM programs which use no Jump instructions (generalize the solutions to parts (a) and (b) of Additional Exercise 2 for this section).

Thus the functions of one variable computable by URM programs with no Jump instructions are precisely those of the forms $n \longmapsto c$ and $n \longmapsto n + c$, where c is a constant.

(b) Suppose that a URM program P contains no Successor instructions. Then the only changes to the numbers in the registers arise from uses of Zero and Copy instructions. A Jump instruction does not change the content of any register. Hence in a computation using P with input n the only numbers that can occur in the registers are 0 and n. If the computation with input 0 stops then the output is 0. For non-zero input n the course of the computation will be the same whatever the value of n so the computation, if it stops, will have output 0 for all n or output n for all n. Thus the only functions of one variable that P might compute are the zero function zero: $n \longmapsto 0$ and the identity function id: $n \longmapsto n$. We already know (Theorem 2.1) that both these functions are URM-computable by programs not using Successor instructions and hence these are precisely the functions of one variable that can be computed by URM programs which include no Successor instructions.

Section 3

1   We can regard this function as obtained by substitution in the multiplication function. However, we do not need to use in full the recipe given in the proof of Theorem 3.1. We do not need to store the input because the program

      1   $S(1)$
      2   $C(1,2)$
      3   $S(2)$

will provide an appropriate input for the multiplication. Also we do not need to use the full program for multiplication given in Example 2.2 because we wish to multiply positive integers. So we don't need to start by checking whether either of $n+1$ or $n+2$ is zero, since this is impossible. An appropriate URM program is as follows.

      1   $S(1)$
      2   $C(1,2)$
      3   $S(2)$
      4   $C(1,3)$
      5   $S(5)$
      6   $J(2,5,12)$
      7   $Z(4)$
      8   $S(3)$
      9   $S(4)$
    10 $J(1,4,5)$
    11 $J(1,1,8)$
    12 $C(3,1)$

2   The function h is given by the equations

$$h(n_1, n_2, 0) = f(n_1, n_2) = n_2,$$
$$h(n_1, n_2, n+1) = g(n_1, n_2, n, h(n_1, n_2, n)) = (n_1 \times n) + n_2 + h(n_1, n_2, n).$$

We then have

$$h(1,5,0) = 5,$$
$$h(1,5,1) = (1 \times 0) + 5 + h(1,5,0)) = 10,$$
$$h(1,5,2) = (1 \times 1) + 5 + h(1,5,1)) = 16,$$

and

$$h(4,2,0) = 2,$$
$$h(4,2,1) = (4 \times 0) + 2 + h(4,2,0)) = 4,$$
$$h(4,2,2) = (4 \times 1) + 2 + h(4,2,1)) = 10,$$
$$h(4,2,3) = (4 \times 2) + 2 + h(4,2,2)) = 20.$$

3 (a) From Theorem 2.1 we know that the identity function can be computed by the following URM program.

$$1 \quad C(1,1)$$

Adapting the program given in Problem 2.2 we find that a URM program to compute the function g is as follows.

```
 1   C(3,1)
 2   Z(2)
 3   Z(3)
 4   J(1,4,11)
 5   S(3)
 6   J(1,3,10)
 7   S(2)
 8   S(3)
 9   J(1,1,6)
10   C(2,1)
```

Now we use the recipe given in the proof of Theorem 3.3, with $k = 1$, $r = 1$, $s = 10$ and $u = 4$, to obtain the following URM program to compute the function h.

```
 1   C(1,5)
 2   C(2,6)
 3   Z(2)
 4   C(1,1)
 5   J(6,7,22)
 6   C(1,3)
 7   C(7,2)
 8   C(5,1)
 9   Z(4)
10   S(7)
11   C(3,1)
12   Z(2)
13   Z(3)
14   J(1,4,21)
15   S(3)
16   J(1,3,20)
17   S(2)
18   S(3)
19   J(1,1,16)
20   C(2,1)
21   J(1,1,5)
```

Notice that, in this program, instruction number 4 (corresponding to the program for the identity function) is redundant and so could have been omitted. In other words, in this context the identity function could have been represented by the *empty program* (i.e. the program with no instructions).

(b) The recursion equations are

$$h(n_1, 0) = n_1,$$

$$h(n_1, n+1) = \begin{cases} 0, & \text{if } h(n_1, n) = 0, \\ h(n_1, n) - 1, & \text{otherwise.} \end{cases}$$

We see that $h(0,0) = 0$ and so $h(0,n) = 0$ for all n. Now suppose that $n_1 > 0$. Then $h(n_1, 0) = n_1$, $h(n_1, 1) = n_1 - 1$, $h(n_1, 2) = n_1 - 2$, ..., $h(n_1, n_1) = 0$ and so $h(n_1, n) = 0$ for all $n > n_1$. Thus in general

$$h(n_1, n) = \begin{cases} n_1 - n, & \text{if } n \le n_1, \\ 0, & \text{otherwise.} \end{cases}$$

4 As suggested in the hint, we adapt the proof of Theorem 3.3.

Instead of two functions of f and g, we have a constant a and a function g. Let B be a URM program with s instructions which computes g. Instead of a program A for f, we are going to need some URM instructions that result in a being in register R_1.

The initial input is n, in register R_1. As in the proof of Theorem 3.3, the first thing we do is to copy n into a register R_{u+1} not otherwise involved in the computations, using the instruction $C(1, u + 1)$.

Next we need some instructions that result in a in R_1. To do this we can clear register R_1 with a $Z(1)$ instruction and follow this by a successive $S(1)$ instructions.

If $a = 0$ then we simply need the instruction $Z(1)$.

We can now select a suitable value for u. Only register R_1 is used in computing a. Let $\rho(B)$ be the largest register number used by the program B. The largest number of registers used for input is 2, in calculating g. Hence we set u equal to the maximum of $\rho(B)$ and 2.

The remaining instructions correspond to those from $k + r + 3$ onwards in the proof of Theorem 3.3 with $k = 0$. The only difference is that in this case there is no need for the Copy instructions of the form $C(u + k, k)$ (i.e. instructions $k + r + 6$ to $2k + r + 5$ in the recipe in the proof of Theorem 3.3 are not needed).

Thus, a suitable program is as follows.

$$
\begin{array}{rl}
1 & C(1, u + 1) \\
2 & Z(1) \\
3 & S(1) \\
\vdots & \vdots \\
a + 2 & S(1) \\
a + 3 & J(u + 1, u + 2, a + u + s + 6) \\
a + 4 & C(1, 2) \\
a + 5 & C(u + 2, 1) \\
 & Z(3, u) \\
a + u + 4 & S(u + 2) \\
 & B \\
a + u + s + 5 & J(1, 1, a + 3)
\end{array}
$$

The initial contents of the registers are

R_1	R_2	R_3	
n	0	0	$\cdots$

Instruction 1 stores n in R_{u+1}. After implementation of instructions $2, 3, \ldots, a + 2$ the contents of the registers are

R_1	R_2		R_u	R_{u+1}	R_{u+2}
a	0	$\cdots$	0	n	0

If $n = 0$, instruction $a + 3$ causes the computation to stop and the output is $a = h(0)$. If $n \neq 0$, there are loops through instructions $a + 3$ to $a + u + s + 5$. Suppose that such a loop starts with register contents

R_1	R_2		R_u	R_{u+1}	R_{u+2}
$h(i-1)$	⋄	$\cdots$	⋄	n	$i-1$

After implementation of instructions $a + 3$ to $a + u + 4$, the contents of the registers are

R_1	R_2	R_3		R_u	R_{u+1}	R_{u+2}
$i-1$	$h(i-1)$	0	$\cdots$	0	n	i

The program B then computes

$$g(i - 1, h(i - 1)) = h(i)$$

and the contents of the registers are

R_1	R_2		R_u	R_{u+1}	R_{u+2}
$h(i)$	⋄	$\cdots$	⋄	n	i

There is then an unconditional jump to instruction $a + 3$. When $i = n$, the computation stops with a jump to a non-existent instruction and the output is $h(n)$ as required.

5   We observe that $2^0 = 1$ and $2^{n+1} = 2 \times 2^n$. Thus the function $h : n \longmapsto 2^n$ is defined by primitive recursion from the constant 1 and the function $g : (n, m) \longmapsto 2m$, since

$$g(n, h(n)) = 2h(n) = 2 \times 2^n = h(n + 1).$$

We can now use the recipe given in the solution to Exercise 4 to obtain a suitable URM program.

A URM program to compute g is as follows.

    1    $J(2, 3, 6)$
    2    $S(4)$
    3    $S(4)$
    4    $S(3)$
    5    $J(1, 1, 1)$
    6    $C(4, 1)$

Following the recipe given in the solution to Exercise 4, with $a = 1$, $s = 6$ and $u = 4$, we obtain the following URM program to compute h.

    1    $C(1, 5)$
    2    $Z(1)$
    3    $S(1)$
    4    $J(5, 6, 17)$
    5    $C(1, 2)$
    6    $C(6, 1)$
    7    $Z(3)$
    8    $Z(4)$
    9    $S(6)$
    10   $J(2, 3, 15)$
    11   $S(4)$
    12   $S(4)$
    13   $S(3)$
    14   $J(1, 1, 10)$
    15   $C(4, 1)$
    16   $J(1, 1, 4)$

INDEX